Every Creeping Thing

True Tales of Faintly Repulsive Wildlife

Henry Holt and Company, Inc.
Publishers since 1866
115 West 18th Street
New York, New York 10011

Henry Holt® is a registered trademark of
Henry Holt and Company, Inc.

Library of Congress Cataloging-in-Publication Data

Conniff, Richard, date.
Every creeping thing: true tales of faintly repulsive wildlife / Richard Conniff;
illustrations by Sally J. Bensusen.—1st American ed.
p.   cm.
Includes bibliographical references (p.   ) and index.
ISBN 0-8050-5697-1 (hb: alk. paper)
1. Vertebrates.   2. Vertebrates—Research.   I. Title.
QL605.C76   1998   98-12700
596—dc21   CIP

Henry Holt books are available for special
promotions and premiums. For details contact:
Director, Special Markets.

First Edition 1998

Designed by Paula R. Szafranski

Printed in the United States of America
All first editions are printed on acid-free paper. ∞
3   5   7   9   10   8   6   4   2

Every Creeping Thing

Richard Conniff

ILLUSTRATIONS BY SALLY J. BENSUSEN

HENRY HOLT AND COMPANY • NEW YORK

Bring forth with thee every living thing that is with thee, of all flesh . . . and of every creeping thing that creepeth upon the earth; that they may . . . be fruitful, and multiply upon the earth.

—GENESIS 8:17

CONTENTS

Contents

ACKNOWLEDGMENTS

I am indebted to many people for their editorial and scientific assistance in helping me bring this book together. I would particularly like to thank Kevin Buckley, who was an editor at *Geo* magazine when we met, for his generous support, and Liz Kristol for her research assistance; at the *Atlantic Monthly*, editors William Whitworth and Cullen Murphy; at *Yankee* magazine, Mel Allen, who first told me about an elusive turtle man rumored to be traveling around New England; at *Smithsonian* magazine, Don Moser and Jim Doherty, also Marlane Liddell, Sally Maran, Bonnie Stutski, Adele Conover, Beth Py-Lieberman, Fran Glennon, Eric Keller, Minna Morse, and Taehee Kim; at *Sports Illustrated,* Bob Brown, Chris Hunt, Margaret Sieck, Connie Tubbs, and Linda Verigan; at *Audubon,* Les Line, Gary Soucie, and Jennifer Reek; at the Discovery Channel, Steve Burns and Mick Kaczorowski; at National Geographic Television, Kevin Krug.

I am especially grateful to Fred Strebeigh of Yale University, whose careful reading and thoughtful suggestions helped give shape to this book; Karen Conniff and James F. Conniff, for help with the manuscript; James C. G. Conniff for editorial fine-tuning; my agent, John Thornton, of the Spieler Agency; my editor at Henry Holt and Company, Ray Roberts; and the illustrator Sally J. Bensusen.

In addition to the researchers and other sources actually named in this book, I also received valuable help from many other people. John Seidensticker of the U.S. National Zoo frequently pointed me toward good stories and sources. I'd also like to thank the following sources and apologize to those I have inadvertently omitted: On "Healthy Terror," J. Walter McNutt for introducing me to the idea of "Pleistocene memories," Rodger Jackman and Lis Leader of Rodger Jackman Productions; on "Days of Torpor, Nights of Sloth," Mel Sunquist of the University of Florida at Gainesville; on "The Cave of the Bats," Chris Fichtel of The Nature Conservancy's Vermont chapter; on "Good Scents," I. Lehr Brisbin of the University of Georgia, William Carr of Beaver College, Stephen MacKenzie of the State University of New York, Larry Myers of Auburn University; on "Looking for Mr. Griz," Paul Schullery of Yellowstone National Park, who first told me about Steve and Marilynn French; on "Cormorant Heaven," J. Brent Harrell of the U.S. Department of Agriculture; on "Acting Like Animals," photographer Jeff Mermelstein; on "A Mouse Like a Spear," Robbie MacDonald of Bristol University; on "Sharks," Wes Pratt, Jack Casey, and Margo Schulze of the National Marine Fisheries Service, Robert Hueter of Mote Marine Laboratory in Sarasota, Florida; on "A Porcupine Would Rather Be Left Alone," the Massachusetts Audubon Society; on "Jungle Days," photographer Randall Hyman, Howard Clark of U.S. Agency for International Development; on "Sleeping with Snapping Turtles," Carl Ernst.

Every Creeping Thing

Introduction:
Healthy Terror

One evening last summer, in a stand of trees beside a floodplain in Botswana, a friend and I were grilling skewered meat over an open fire. A pride of sixteen lions was living somewhere off to the east of our camp, and another pride of twenty-two lions to the west, and one night a leopard wandered past, making a low growling sound like a ripsaw slowly notching through somebody's thighbone. An elephant had lately been chasing people in the neighborhood, with a high, outraged trumpeting that I could hear with the hairs on the back of my neck. It was of course unnerving to live among all this wildlife, protected at night by nothing more than a layer of ripstop nylon, but it was also exhilarating. Looking out across the grassy floodplain, where ten-foot-tall termite mounds pointed like monumental fingers to the sky, I could imagine what it might have been like to be a hunter-gatherer at a time not so far in the past, anywhere in the world, in a wilderness now surely vanished.

We squatted by the fire watching the flames lick at the charred corners of the meat (it was pork satay, lest this all seem a little too primitive), and the conversation turned to what my friend termed "Pleistocene memories." He'd heard a lecture once theorizing about the evolutionary roots of our aesthetic sensibilities. We love lawns, to cite the most familiar example, apparently because short grass made it easier for our forebears to see snakes approaching. We love clean, polished surfaces because they suggest the proximity of water. (This same touching confusion may also trouble animals. Like us, female dragonflies sometimes get shiny cars and reproduction mixed up: They hover expectantly over a car and fire their eggs down onto the polished hood, mistaking it for the surface of a pond.) We tend to respond positively to a certain branchy tree shape because it stirs up comforting genetic memories of the sort of tree in which our simian forebears roosted at night, for safety from predators.

The embers gyred up toward the treetops, and with them went my thoughts, in which the lives of people and animals were twined together in an endlessly varied dance of fear, worship, curiosity, delight. Among other things, I found myself thinking about Gaius Julius Solinus.

Regarding the third or possibly fourth most influential writer of natural history ever produced by Western civilization, almost nothing is known except that he was a systematic plagiarist and suffered from an inordinate fondness for monsters. Classicists have generally surmised from close readings of the text that Gaius Julius Solinus was a Roman, or anyway a citizen of the Roman Empire, who cobbled together his *Collectanea rerum memorabilium* (*Collection of Remarkable Facts*) sometime between A.D. 220 and 280. It was a work of about two hundred pages, mostly stolen from the Roman encyclopedist Pliny, and when modern critics try to characterize it, words like "wretched" and "trivial" spring to mind. It is a source of understandable consternation for these critics to add that Solinus nonetheless helped shape European ideas about nature and the world at large for more than a thousand years.

So when I came across his work not long ago, in the course of preparing this book, Solinus naturally intrigued me. I browsed through his remarkable facts with wonder at how shamelessly he distorted the natural world. The deeply gratifying sensation of disdain for our primitive forebears welled up within me; here was the perfect model of everything a natural history writer should not be. But I must confess that I also continued to read, and with a sort of guilty fascination otherwise reserved for tabloid newspaper headlines in the supermarket checkout line. You know the sort of thing: "40-Foot Moth Crashes Into Lighthouse—Monster bug couldn't resist glowing beacon."

As the title of his book promised, Solinus presented readers with a world of marvels, some pleasing, but mostly horrific: In Germany, the nocturnal traveler could depend on luminous birds to light his way through the forest. In Ethiopia (a vague term referring to any place from Senegal to Sri Lanka) lived ants the size of mastiffs that used their claws for digging gold. Egypt was subject to periodic invasion by vast flights of winged serpents, against which the vigilant ibis was its only defense. In India dwelt the cruel manticore, a lion with a scorpion's tail and the face of a human, which "seeketh most greedilie after man's flesh."

The natural world as Solinus presented it was deeply polarized. Some animals were intuitively sensitive to human needs and properly submissive in serving them, while others, like the manticore, lived chiefly to rip human flesh—which also took strange and monstrous forms. Despite the utterly anthropocentric character of the world as Solinus saw it, the boundary between humans and animals was confused, as it remains for us today. Solinus populated distant nations with Cynocephali, or dog-headed humans, and with headless people called Blemmies, whose faces were incorporated into their chests.

Such marvels no doubt found an eager audience in part because of their storybook quality; in time, Solinus and his book alike became known simply as *Polyhistor,* the teller of many tales.

Manticore

But unlikely as it may seem to modern readers, Solinus endured for a millennium because people also believed what he wrote, and they wanted to believe it. Strange creatures out of Solinus thus turn up repeatedly not just in fables but on sober world maps, and not just through the medieval period but well into the age of exploration. He provided the stock images to which the lazier map illustrators returned even into the 1700s, as in the Jonathan Swift verse: "So Geographers, in Affric-maps, / With savage pictures fill their gaps . . ." One such stock image—found, for instance, in the *Nuremberg Chronicles* of 1493—were the Sciopods (or "shade foots"), a swift but single-footed race of humans who dealt with the midday equatorial sun by flopping onto their backs and raising their huge feet overhead as parasols. When early explorers did not actually find Sciopods or manticores where Solinus and subsequent writers put them, wishful mapmakers simply moved these figures farther out into terra incognita. Thus did Solinus persist, on the strength of what Rabelais later characterized as "beautiful lies."

He persisted, in part, I think, because he tapped a genuine and badly underrated human emotion about the natural world. Solinus understood that nature scares a lot of us silly—and this is one reason he caught my attention, not just that evening squatting by the fire, but afterward, as I thought about what I hoped to accomplish with this book: Solinus seemed to me to be a useful antidote, albeit an extreme one, to the fuzzier one-with-the-universe attitudes of our day.

Let me be more explicit about this, at the risk of saying good things about a very bad book, and incidentally of appearing to be a throwback. Despite the efforts of many earnest and life-affirming people to persuade me that the vampire bat is our friend and that Native Americans enjoyed true harmony with Brother Wolf, I have never quite overcome the gut feeling that fear of nature is normal—more normal, certainly, than the love of it. Or perhaps I should say that fear and love are thoroughly tangled together. We have evolved over hundreds of thousands of years as hunter-gatherers, not as office workers, and our genes are still encoded with all the alarm signals that were appropriate when we lived among lions. We like to pretend these feelings don't exist, or that they are leftovers from so far back in our collective history as to have dwindled to imperceptible faintness. But even in Europe, *Homo sapiens* dwelt side by side with lions until about A.D. 200. Hippos once lived as far north as England, and rhinos appear in the cave paintings at Lascaux, on the border of France and Spain, from 12,000 B.C. Only a few generations ago, grizzly bears ranged from California to Ohio, and wolves roamed on Cape Cod. In the course of my research for this book, I once hiked out into an open meadow in Yellowstone National Park looking for grizzly bears. Instead, I wandered too close to a small herd of impassive bison, which I regarded with indifference until one of them took offense, glowered, and came rolling toward me. It was less a charge than a warning, but, oh, how my Pleistocene heart surged.

In his book *The Way of the Animal Powers,* Joseph Campbell describes how, even after generations of laboratory rearing, newly hatched chickens run for shelter if a hawk passes overhead, but they ignore gulls and other unthreatening birds. He wonders if such "archaic" genetic codes also sleep within us from our long evolution among woolly mammoths and cave bears. We react with alarm, for instance, to the mere shape of snakes or spiders, though it may have been thousands of years since either represented a serious threat in our daily lives. Campbell quotes Wordsworth on the origin of our souls in God, from whom we arrive "trailing clouds of glory." But the paleontological evidence is that "our life star's other setting" was also earthly and left us trailing clouds of fear.

The fear of nature is not only normal; it can also be pleasurable. One reason I spend much of my time writing about the natural world, and particularly about the animals in this book, is that nature gives me the creeps, and the more I learn the creepier and more wonderful it gets. There really are plenty of terrible things out there to tremble about—not winged serpents and manticores, of course, and not even grizzly bears or great white sharks either, at least not in the conventional sense. (No, I'd better take that back. Any time I dip a toe in the habitat of either one, I feel a tingling awareness that this might just be the one fluky time a human gets ripped to little sushi strips.) What I really find creepy and wonderful about nature are not its great terrors, but its weird, unsuspected minutiae, its intricate workings painstakingly developed over millions of years of evolution, its minor horrors— for instance, that some sharks practice sibling cannibalism in the womb, or that a mole will paralyze earthworms, ball them up in a knot, and seal them away in individual cells in the walls of its chambered mound, still living, to be eaten at leisure. I am captivated by the sight of a keyhole limpet on the rocks off British Columbia being attacked by a sea star. The limpet protects itself by lifting its mantle up over its shell, exposing a smooth surface on which the sea star can't get a grip. But the limpet also carries a sort of vicious pet under its shell, like an old lady's lap dog, and if the sea star persists, this pet, actually a small worm, reaches out and nips chunks of flesh from the assailant's feet. Nature fascinates and horrifies me in equal and directly proportionate measure, and I like to keep a respectful distance.

It seems to me that the modern world, in full retreat from the fang-and-claw nightmares of the past, tries too hard to do the opposite, to be cool about the natural world when our hearts are quaking, to exhibit healthy scientific attitudes, like kids in a junior high school sex education class. I think of David Attenborough purveying ecological dogma sotto voce while very nearly sitting in a gorilla's lap, or of a woman I once watched pluck a tree boa from a branch in a Peruvian rain forest and wrap it around her forearm as if it were a bracelet from Bloomingdale's. There is something a little patronizing in this commonplace swim-with-the-dolphins mentality. We are in danger of turning what's left of the planet into a petting zoo.

Aldous Huxley once wrote that nineteenth-century English romanticism about nature was possible only because the dark and terrifying forests had been cut down, and the landscape opened up. All those broadaxes made the Lake District safe for Wordsworth. Similarly, our benevolence toward wilderness, while appropriate to our times, is a measure of how much more thoroughly the wilderness has been subdued, and of our urbanized distance from it.

The catalogue of horrors put forward by Solinus is, in contrast, a testimony to nature's strength. One reads it with a sense of nostalgia for the sound of wilderness still scratching and howling just outside Europe's garden gate. Solinus reminded me that what is missing from our attitude toward nature is a modicum of healthy terror.

That said, it must immediately be added that, as a writer about natural history, Gaius Julius Solinus was an utter booby, the embodiment of everything the worst popular writers about natural history have ever done wrong. His unfailing instinct was to reach for the best story, rather than sifting through the lore for something like truth. Even in Italy itself, Solinus apparently did not fear that bane of "beautiful lies," the skeptical and inquiring reader. In Calabria, he wrote, boas sucked milk from cows and laid waste the countryside. Italian wolves were endowed with a very fine tail hair "that hath the power of love in it," if only the ardent youth could pluck one from the living animal. The urine of the Italian lynx congealed into an amber stone (dubbed "lynx-pisse" by the sixteenth-century English translator), which was endowed with magnetic powers. But the lynx was so spiteful it swept dirt over its leavings to prevent them from being turned to human profit.

Solinus appears never to have wasted a minute studying a real animal or plant. He was a grammarian. He studied texts, and particularly one text, Pliny's thirty-seven-volume *Natural History*. In his *Collection of Remarkable Facts,* Solinus made 1,150 unacknowledged borrowings from Pliny, a prodigious average of a half-dozen plagiarisms per page; hence his nickname "Pliny's Ape." His program, in one critic's scathing view, was "to extract the dross and leave the gold" from Pliny's "not altogether unscientific" work, producing "the most completely miraculous view of the world ever put forward in Europe." The pity in this was that Pliny himself was an eager student of nature; so much so that he died from the fumes while observing the eruption of Mount Vesuvius at close range.

So if I share with Solinus the universal human fear of nature, and also a fondness for its remarkable facts, I have labored to be sure my facts are real, and wherever possible, I have gone out to see them for myself in the field. In the course of writing this book, I have traveled to observe different species in England, Scotland, Ireland, Ecuador, Panama, Peru, Bermuda, the Bahamas, and, among other U.S. locations, California, Connecticut, Florida, Maine, New York, Vermont, and Wyoming. I have gone to these places not just to see the animals, but also to meet the people who know them best, a mix of biologists, hunters, hobbyists, animal trainers, and conservationists. For me, these travels have been a great pleasure, occasionally uncomfortable but almost always enlightening. The people who actually spend their lives in the field, observing the animals, would probably say the same thing for themselves. But, like Pliny, they seem often to have paid a heavy price for their love of nature, including meager salaries, minimal professional recognition (academic advancement occurs in inverse proportion to time spent in the field), troubled marriages, and such colorful medical disorders as malaria, encephalitis, bubonic plague, filariasis, leishmaniasis, hepatitis, botflies, arboviruses, giardiases, worms, amoebas, and assorted bites and stings. One of the researchers described in this book was suffering from an unidentified tropical virus when I visited with him, and has since dropped out of biological research. Another saw his wife killed in a fiery accident at their field station. Two researchers, each carrying in his mind an irreplaceable library of knowledge about the natural world, flew into a cloud forest mountaintop, and died.

In choosing to write about these people, and to see the animals as much through their eyes as my own, I have, like Solinus, written an anthropocentric book. Purists may object that the lives of the animals merit consideration entirely on their own, and I would not disagree with them. For instance, the improbable discovery that five-hundred-pound grizzly bears choose to spend the month of August in the mountains of Wyoming, feeding almost exclusively on moths, gets us nowhere, from a practical point of view. It is merely a remarkable fact, and the idea that the natural world is full of a million such lovely, paradoxical facts, most of them still undiscovered, or even unimaginable, ought to be sufficient. We shouldn't need to argue for preserving grizzly habitat, because grizzlies produce economic benefits through ecotourism, or because grizzlies may contain something in their genetic code that may some day, somehow, prove useful for human survival. A sense of wonder at the strange workings of evolution should be enough.

But there is also room, it seems to me, to wonder at the strange workings of our own engagement with the natural world. Beneath the modern scientific attitude toward nature, our undercurrent of lore about animals runs almost as strong as it did for Solinus, or for the Neolithic hunters who painted the murals by torchlight in the cave at Lascaux. About the group of weasels known as stoats, for example, I have been enthusiastically informed that they suck blood, can hypnotize rabbits, carry their dead home in a funeral cortege, and sometimes conceive a grudge against a human so strong that they will scrabble against the window glass in a vampirish frenzy to get at the victim within.

It is unwise to be smug about Solinus because people are always going to need the kind of lore about the natural world, the mix of wonder and dread, attraction and repulsion, that he provided. It is human nature—and I don't mean that to convey any hint of regret. Animals are the source of some of our most imaginative and persistent fantasies, and these fantasies are the only way most of us ever get to escape from our urbanized and domesticated lives into a larger world. Not long ago, I read a Seamus Heaney poem in which he accomplishes the miraculous feat of writing a tender love poem to his wife by likening her to a skunk. It is a poem about a man on the road in California, lonely for home, visited at night by a skunk in the backyard:

> And there she was, the intent and glamorous,
> Ordinary, mysterious skunk,
> Mythologized, demythologized,
> Snuffing the boards five feet beyond me.

The writer eventually goes home and forgets the skunk, until, as he tells his wife:

> It all came back to me last night, stirred
> By the sootfall of your things at bedtime,

Introduction: Healthy Terror

Your head-down, tail-up hunt in a bottom drawer
For the black plunge-line nightdress.

It is surely an eccentric poem, but also lovely, and it reminded me how much we need to keep the threads of tenderness and savagery alive in our vision of the natural world.

These threads run with special vividness through the lives of the people I have chosen to write about, though they might perhaps deny it. The scientific researchers in particular work under a burdensome obligation to regard nature with clinical dryness, devoid of emotion or anthropomorphism. By inclination, most of the characters in this book are also down-to-earth; they would say they are just trying to make a living, muddle through, get by. Ted Parker, one of the most remarkable ornithologists of this century, once told me that he started to memorize birdsongs as a child simply because it was a way to impress the little old ladies in his Lancaster, Pennsylvania, birding club. But by the time I met him, in his mid-thirties, he had memorized the songs of more than four thousand neotropical bird species, and often knew multiple calls for each species. Birdsong was a language for him. Shamans on their mystical bird flights into the otherworld could have no more profound a connection with the animal spirits. Much as he might claim to be just a guy in high-top Converse sneakers looking for a pickup basketball game, it was clear that Parker had been singled out and anointed by Nature—albeit in the person of a few elderly women in a Pennsylvania suburb. Likewise, Nature could hardly have been more explicit in choosing David Wingate, the Bermuda conservationist: When he was just twelve years old, a tidal wave trapped him and his dog on a sandy hillock, "and I realized, 'Christ, I'm on an island,' and then the water receded."

The darker threads of our engagement with the natural world also run through the lives of everyone in this book. There is, for instance, no one who loves snapping turtles or identifies with them as much as does John Rogers. But he is a turtle hunter by profession: "I spend the whole week with turtles. It's just me and them, me and the snappers against the world. And then what do I do? I take 'em into the cities and the suburbs and I sell 'em. I'm selling my friends . . . and it's always the same feeling . . . the truck's going to be empty and I'll be alone again."

The flip side of healthy terror, a primordial loathing for things that are different from us, a powerful impulse to destroy the encroaching natural world, can at times seize hold of anyone. Once, over too many large *cervezas* on some backwater of the Amazon, a biologist told me that, as a boy, he used to make popsicle-stick constructions. But they weren't rafts or other commonplace playthings. Instead, he created an upright framework, in which he mounted a single-edge razor blade with a lead weight on top, creating a guillotine, through which he put any snake he could catch. As an adult, perhaps to compensate, he was making a living by defending animals from inhumane treatment. His life represented an extreme transformation. But we all have such shameful memories (I can still hear the sound of a turtle's skull that I crushed under a stone as a child), and if we never face the dark side of our connection to the natural world, we will never understand nature itself.

In the end, my intention has been to see both people and animals as they are, not to homogenize them, not to make them merely nice, not, above all, to turn animals into petting zoo creatures so we can feel good about ourselves when we protect them. Which brings me back to Solinus. If his *Collection of Remarkable Facts* plainly wasn't looking at the real world, neither were his readers over the next millennium, and this disinclination to question, to explore, to rely on what their own senses might have told them, may be the hardest thing for modern readers to fathom. How did accepted wisdom triumph so thoroughly over experiment, over experience, even over reality?

Roman writers of Solinus's day typically tried to cite as many notable authors as possible, and they bequeathed to the Middle

Ages the naive idea that the one "who could assemble the greatest number of authorities on a subject was himself the most reliable authority." The apparent comprehensiveness of all those endlessly cited sources, the sheer pedantic deadweight, smothered any urge to do original research. "For what can we challenge properly for our own," wrote Solinus, with a hapless intellectual shrug, "since the diligence of menne in olde tyme hath been such, that nothyng hath remayned untouched unto our dayes?"

Through this sort of uncritical repetition, European knowledge about the natural world underwent a transformation that T. H. White has likened to the nursery game Russian Telephones, in which children whisper a message from ear to ear, and what started out as "Nanny thinks this game is fun" comes out in the end as "Granny wants an Uzi gun." Thus one Roman moralizer mistranslated from a Greek original and ascribed to the shark (*galeos*) the supposed reproductive habits of the weasel (*gale*). Real animals were beside the point as the game of Russian Telephones proceeded through the Middle Ages.

After Solinus, almost all of the players were Christians. The Church, with its developing preference for dogma rather than individual perception, readily took up the Roman regard for authority, even certain pagan authorities like Solinus. But where Solinus distorted reality in the cause of entertainment, the Church distorted it as a way of disclosing Christian truth. What mattered, Saint Augustine wrote, wasn't whether certain animals actually existed, so much as what they *signified*. Thus "the most completely miraculous view of the world ever put forward in Europe" found its ideal readership. The Western world decided that it could get by with a grab bag of ideas about nature and for centuries turned its back on nature itself.

All this is worth thinking about now because the world seems once again to be turning its back on reality, shutting nature out. We want to preserve our animals, but only in isolation, separate from our everyday lives. We want rain forests as a day trip from a first-class hotel. We want rampant elephants as the raw material for a Discovery Channel documentary. But Kenya has proposed fencing in its elephants, and in Botswana's last great wilderness, the Okavango Delta, where I squatted by the fire and contemplated Pleistocene memories, more than nine hundred miles of cattle fencing have been erected in the past few years, cutting off old migration routes and killing wildebeests and giraffes. The dull generalists of the animal world—the mallards and mute swans—are everywhere in the ascendant. The stranger, more idiosyncratic species are vanishing, or locked up in zoos. We have no shortage of entertaining new mythologies to substitute for this bland reality—from Tamagotchi cyberpets to the extraterrestrials in endless movies of the *Mars Attacks* genre. In coastal Ecuador, where Ted Parker and I were watching one of the last precious remnants of rain forest being cut down around us, the landowner visited one day to check on the progress of the chainsaws. His young son came with him, wearing a Teenage Mutant Ninja Turtles T-shirt. Raphael, Donatello, Leonardo, and Michelangelo lived, and the kid was undoubtedly oblivious to the far more bizarre, more combative, more *radical* life being erased forever before his eyes. It's possible that that unfortunate kid, and the rest of us, can get by with such substitutes, much as the Middle Ages got by with Solinus.

But we shouldn't have to: We are fortunate to live in an age of discovery. Because of people like Parker and Wingate and many others in this book, we know that the world is full of life-forms that are too strange, too beautiful, too powerfully disturbing to let merely vanish. We can celebrate—before it's too late—all the strange and wonderful things biologists are learning about how the natural world really works, and all the oddball creatures they continue to discover. We can actually choose between wilderness and the petting-zoo world. We can accept an artificial future in which "all creatures great and small"—or the prettier ones, anyway—are preserved and safely packaged for tourists and television viewers. Or we open the gates and stand back, as God ordered Noah to do in Genesis: "Bring forth with thee every living thing that is with thee, of all flesh . . . and of every creeping thing that creepeth upon the earth; that they may . . . be fruitful, and multiply upon the earth."

We can preserve for ourselves, and for its own sake, a world of remarkable facts as entertaining as anything in Solinus—indeed, more compelling than anything he could ever have imagined. Only this time the creatures are real. If our vision of this world is good enough, maybe the third millennium, unlike the first, will not have to make do on fantasy alone.

Days of Torpor,
Nights of Sloth

The first time I saw a sloth, in a small and sparsely furnished cage at the National Zoo in Washington, D.C., it took me several moments to spot the thing: a motionless blob of fur clinging to a bare tree branch.

"It's a fake," a teenager advised me. "Fake something or other. There are two of them."

"They're gorillas," another teenager ventured. "Little baby gorillas."

"Then how come they don't move?" said the first teenager.

The second teenager consulted the small sign posted in a corner. "Maybe it's a leaf-nosed bat," he said, reading the name of a species said to be in residence. But the idea of a bat the size of a small dog (and leaf-nosed to boot) was so unsettling that after a brief pause all of us descended as one on the sign. "Oh," said the second teenager, "they're sloths."

"What? Those big ugly things?"

"Yeah. It says right here. Two-toed sloths."

Full of doubt, his companion replied, "I thought sloths were supposed to be extinct."

Pity the sloth. It is an animal perpetually misunderstood and reviled by the human species. These inoffensive mammals, which pass most of their lives thirty to ninety feet up in the rain forests of Central and South America, have been called "normal morons," "chronic pacifists," and even "hanging animal baskets." Georges-Louis Buffon, the eighteenth-century naturalist, listed among their chief attributes "slowness, stupidity and habitual pain," and added that with even a single additional defect, sloths would cease to exist. Indeed, the sloth's very name is a deadly sin. Says the Bible: "Slothfulness casteth into a deep sleep; and an idle soul shall suffer hunger."

But there is a paradox here. The sloth truly enjoys a deep sleep (ten to fourteen hours a day), and it is by definition idle. Some sloths are so lethargic as to be utterly unmoved by gunshots at close range or by the presence of cats, eagles, or other common predators. And yet sloths survive, lazing in the upper branches of the jungle, while quicker, keener, more industrious animals around them fall victim in the daily struggle. As to hunger, the sloth's stomach is disproportionately large. Together with its contents, it may make up a quarter to a third of the animal's weight. And it is usually full. Sloths have the good sense to survive largely on leaves, and the supply is abundant. They do not often suffer hunger. The sloth is, in short, an affront to all our notions about the predatory jungle, and about life itself.

Shortly after my zoo visit, in search of a somewhat more informed point of view on these enigmatic creatures, I traveled to one of the places they call home, Barro Colorado Island in Panama. There are probably ten or more sloths in every hectare (or four or more per acre) on this island reserve, which is maintained by the Smithsonian Tropical Research Institute. But naturalists may work for months without noticing one. An exception is Bonifacio de Leon, known to other naturalists on the island as "Boney," who had been described to me as possessing an almost supernatural ability to locate sloths. I found him at his workbench, with the skulls of monkeys and three-toed sloths hanging from monofilament overhead. He agreed to take me on his morning tour, and after he'd put on a silver hardhat and taped his pant legs shut at the ankles as a precaution against ticks, we set off.

Five yards into the forest, he stopped, pointed, and said, "*Perezoso*" (Panamanians call the sloth *gato perezoso,* or "lazy cat"). When it became apparent that I could see nothing, Boney seized me by the shoulders and fixed me at the proper angle of vision. Still nothing. He plunged into the undergrowth, and I followed him fifty feet in, past a dozen substantial intervening trees. Just as he was about to start climbing, I discerned a three-toed sloth sixty feet up. At that moment, it was scratching its butt, rather ineffectively.

Back on the trail, Boney pointed out an assortment of wildlife, with which he communicated by sucking noisily on his wrist, by smacking his lips against his fist, and by humming on a torn off piece of leaf. Some of the creatures he addressed were visible—a parrot, a toucan, an anteater. Some were merely audible—the capuchin monkeys crashing through the treetops. A few—the sloths—seemed neither visible nor audible. Boney helped me in my obtuseness. When I failed to perceive a two-toed sloth sleeping fifty feet up in a hammock of vines, he picked out two vines hanging down from the canopy and hauled on them like a bell ringer in a church tower. The sloth stirred and became apparent, moving off like an upside-down cat that is slightly peeved at being disturbed. "*Perezoso,*" Boney explained.

I consoled myself with the thought that even Gene Montgomery, then a vertebrate ecologist at the Smithsonian Tropical Research Institute, had a hard time spotting sloths. Throughout the 1970s, Montgomery and a colleague, Mel Sunquist, conducted the first methodical study of sloths in their native environment. The two of them, and sometimes Boney, too, climbed trees to capture animals for observation, fed glass beads to captives to measure digestive flow, strapped radio collars on free-roaming individuals, tucked minitransmitters up the animals' backsides, and otherwise arrived at a detailed knowledge of the sloths of Barro Colorado Island.

Studying sloths requires patience above all (watching them in action has been likened to reading *War and Peace*), but in Montgomery I did not find the phlegmatic, professorial "Dr. Sloth" I had been expecting. He arrived to meet me outside his office in Panama City at the wheel of a battered four-wheel-drive vehicle. Solidly built and ruddy-skinned, he wore a great walrus moustache and a characteristic facial expression of roguish glee. The first point he wanted to make about sloths—almost before we sat down—was that they are "nice animals." Or rather, *three-toed* sloths are nice, anyway.

The distinction matters, as he soon explained. The five species of sloth fall into two genera, whose scientific names are *Bradypus* (which means slow-footed) and *Choloepus* (which means crippled). Montgomery pulled out a cardboard box full of animal bones in plastic bags. "That one's an anteater," he said, throwing a bag back and continuing his search. When he'd found suitable specimens, he demonstrated that the two sloth genera can be distinguished by the number of hooklike claws on their forelimbs; hence the common names two- and three-toed sloths. They are as different, he said, as cats and dogs.

Two-toed sloths, like the blobs of fur I'd seen in the zoo, are far and away the livelier genus. Compared with other mammals, two-toed sloths may seem dazed or drugged. But three-toed sloths in action can be compared only with plants, and even then, kudzu may be quicker. They move as if put into a trance and then filmed in slow motion. Indeed, Montgomery's predecessors in sloth research did not require radio telemetry. To test the mobility of the three-toed sloth in the wild, naturalist Hermann Tirler placed a plastic bowl on an animal's head one night. It was still there in the morning. Another time, he briefly kept a three-toed sloth as a pet: "One evening," he wrote, "we suddenly smelled something like a burning sloth, the odor coming from the neighboring room." The animal was half-asleep on a large electric bulb, with its rear going up in smoke, and according to Tirler, it wanted to stay there.

While climbing ninety-foot-tall trees, some of them too thick for a safety belt, Montgomery understandably came to prize the three-toed sloth for its immobility. It was a relatively easy matter to reach out with a noose at the end of an extension pole, pluck the animal from its perch, and lower it to the ground. Close encounters at that altitude were also not unpleasant. Even if a three-toed sloth were to rouse itself to attack, the only creature

Three-Toed Sloth

slow enough to suffer its bite would be another three-toed sloth. In any case, said Montgomery, it has no incisors and its bite is no worse than a somewhat leathery kiss.

The two-toed sloth, on the other hand, has sharp canine teeth and can make itself very unpleasant when threatened. Hissing, it swings an arm out to hook an adversary and draw it in toward its mouth. Montgomery showed me a knuckle on his hand that was once cut to the bone in this fashion. As if being bitten by a sloth were not embarrassing enough, he noted with chagrin that the animal was under sedation at the time. The two-toed sloth is also quick enough to run away when provoked. Montgomery recalled afternoons in which he climbed one ninety-foot tree only to have his quarry escape to the next tree over, then climbed that tree only to have the sloth escape to a third tree, and so on until "the air was fairly blue with strong language." Radio-tracking showed that two-toed sloths also do a considerable amount of traveling through the trees without provocation. Unlike three-toed sloths, they are strictly nocturnal, and on their nightly ramblings they may use their canine teeth to raid birds' nests; in captivity they also eat fruits and vegetables.

I suggested to Montgomery that if one compared the two kinds of sloths it might seem that the three-toed genus is more primitive, a less-advanced life-form. (Not only is its lifestyle one of more profound torpor, but as Buffon suggested, it seems fragile, on the verge of extinction. No one has ever managed to keep a three-toed sloth alive in captivity on any diet.) Montgomery regarded my suggestion balefully. On the contrary, he said, it was possible to argue that the more active genus is, in a sense, less perfectly adapted to its environment. The lazier, less aggressive three-toed sloth has advanced further into the sweet ecological niche of slothfulness. Every feature of these animals has evolved to make a leaf-eating, tree-dwelling life easier.

Besides being hard to reach, leaves are also difficult to break down into useful nutrients; their single virtue is their abundance.

To exploit this resource, both two- and three-toed sloths had to make compromises, and it might well be said that they gave up everything else in exchange for a permanent meal ticket.

Muscle, for example. Maintaining muscle bulk requires more energy than a sloth can easily extract from leaves, so it gets by on about half the usual amount. The two-toed sloth weighs about ten pounds, the three-toed just seven, and only about a quarter of their body weight is muscle. Low body weight is an advantage; it allows the sloth to climb out on the slenderest branches, where it can harvest leaves at a leisurely pace, safely beyond the reach of most predators. But lack of muscle is also the reason sloths are so slothful. They simply don't have what it takes to go any faster. On the ground, where they seldom travel, sloths drag themselves along on their bellies. I saw several animals attempting this in captivity, on a concrete surface, and they looked like men dying in the desert just short of the water hole. But hanging under a branch requires little energy, especially as sloths can lock their hooklike claws in place and concentrate on the more important business of sleeping and eating. Their grip is so tenacious that, according to myth, they remain hanging in place even after death. At times, for want of an easier way to bring down a sloth, Montgomery had to saw off the branch to which the animal was attached. South American Indians who eat sloths sometimes bring them home in this fashion, with the branch over a shoulder and the sloth hanging off the end, riding indolently to its doom.

Immobility is otherwise a virtue for sloths. Their stillness makes them harder to see, and not being seen is really their best form of defense. All their extremities are adapted to crop the maximum area of surrounding leaves with the least possible movement. For example, said Montgomery, other mammals, even giraffes, have seven vertebrae in their necks. He showed me the skeleton of a three-toed sloth; it had nine. The three-toed sloth can bury its pestlelike head and neck in its chest or swing it back perpendicular to the spine to get at leaves behind its head. It can also rotate its head through three-quarters of a circle, and it does so regularly. One of the sloths Boney showed me on Barro Colorado kept looking slowly from side to side like a bewildered child in bottle-bottom glasses. Having settled into a verdant spot, with food all around, sloths do not, however, gorge themselves. They masticate slowly and, according to one source, take in only about one-seventh as much sustenance as a young fawn of the same weight.

In its quest to remain unnoticed, the sloth benefits not just from its immobility but also from its excellent natural camouflage. As my instinct for sloths became keener during my stay on Barro Colorado, I spotted one sitting in the crotch of a tree, arms folded across its chest, patiently waiting for nearby buds to burgeon. It resembled nothing more animate than a nest or a bunch of dead leaves. But as if this were not enough, I noticed that its pelt had a greenish tinge that blended in with the foliage. Montgomery later explained that three species of algae grow in its grooved, gray-brown hairs.

After immobility, the sloth's other great adaptation to leafy life is its gut. The eighteenth-century Irish author Oliver Goldsmith described the sloth as "the meanest and most ill-formed of all those animals that chew the cud." In fact, chewing the cud would take far too much energy to be of any value to a sloth. Instead, it allows the bacteria in its multichambered, ruminant-like stomach to do the work of breaking down cellulose into energy. No other animal devotes so great a portion of its bulk to digestion. One researcher described the sloth to me as simply "a big chemical factory, a fermentation chamber that moves around in the trees."

Despite the size of its stomach, the sloth digests (need one say the word?) slowly. It is a very regular animal. Montgomery and Sunquist found that three-toed sloths defecate and urinate approximately once every eight days, two-toed sloths slightly more often. This lethargic digestion may be due to the animal's inability to maintain a constant body temperature. It ranges from a high of 91 degrees Fahrenheit to as little as 75, and when the temperature drops to its daily low point, the chemical factory may simply shut down.

The sloth's toilet habits are remarkable. Its excreta are almost odorless and would probably not betray the animal's presence if merely dropped to the ground from fifty feet up. But when nature calls, the sloth always descends laboriously to the base of a tree. Hanging on to a vine with its forelimbs, in a position of extreme vulnerability to predators, the three-toed sloth actually digs a shallow hole with its stubby tail before emptying bowels and bladder. It then covers up its wastes with dead leaves, and climbs slowly back up into the canopy.

What makes this ritual even more bizarre is that it is the focus of intense interest among a vast group of specialized insects living in the sloth's hair. Sloths harbor nine species of moths, four species of beetles, six species of ticks, and other assorted mites, and receive frequent visits from itinerant mosquitoes and biting sand flies—altogether constituting a remarkable instance of the profusion and interdependence of tropical life. Researchers have gathered as many as 120 moths from a single sloth, and a record of 978 beetles, all of them waiting patiently for the sloth's weekly defecation, which is both a source of nutrition and an ideal site for egg laying. This population might seem even lazier and less demanding than the sloths on which they live. Adult moths, for example, sometimes break off their wings in the host's fur and remain contentedly sloth-bound for life. But the competitiveness among tropical fauna is such that some of the beetles and mites actually enter the sloth's anus to lay their eggs, gaining several days' incubation time on their less intrepid rivals. All this the sloth bears with equanimity.

The three-toed sloth spends several hours a day grooming, but with little effect on beetles or moths, which can be seen advancing in waves just ahead of the sloth's slowly probing claw. Montgomery believes the grooming is mainly intended to fluff up the sloth's hair and an underlying layer of fur for better heat retention.

The sloth is an excellent swimmer, and this, too, might be expected to have a discouraging effect on its insect friends. The leaves crammed into its gut give the sloth excellent buoyancy. I saw sloths swim more than a mile across Lake Gatun and the Panama Canal to get to and from Barro Colorado, and they manage both a creditable breaststroke and a passable backstroke. The French naturalist Marcel Goffart was moved by the same sight to rhapsodize: "Their arms work strongly, and their behind waggles elegantly. Gone is their sluggishness. . . . Gone is the usual laziness. The swimming sloth is beautifully easygoing." As to insects, the dousing does not bother them. They crowd up in the dry hair of the sloth's head and back, waiting out the trip like passengers on the Friday night ferry to Nantucket. (Scientists can only speculate about why an arboreal mammal would have become such a good swimmer. But it is obviously a useful skill in flood-prone areas and also helps as a dispersal mechanism.)

From swimming, we come to sex. Goffart presents sloth lovemaking as a sort of splendor in the jungle. He reports hearing about one affair that lasted two days, and of a second encounter, he writes: "They were locked in close embrace, *ventre à ventre*. No signs of fighting were evident. Several general muscle spasms were said to have taken place over a period of half an hour, not the least attention being paid to onlookers."

Alas, the only reliable eyewitness account comes from the National Zoo in Washington, D.C., and it is considerably less romantic. Sloths do, in fact, have sex face-to-face. Like humans, they form the "two-backed beast," and they do so while hanging from a branch. The female assumes the sloth's characteristic inverted position. The male, also inverted, works his way between her and the branch. Then he turns over until they are face-to-face, with the female becoming a sort of hammock. Here is one instance where humans might wish the sloth to be truly slow. But in the encounter witnessed by Larry Newman of the National Zoo, the male sloth made about ten misguided thrusts, finally achieved penetration, then finished off with three or four quick thrusts before withdrawing. Newman does not say whether the male fell asleep on the spot or at a polite distance. The female gave birth eleven months later.

The sloth's strategy of a long, leisurely life (up to 40 years), and a low rate of reproduction has made it a remarkable success from Honduras and Nicaragua in the north to about São Paulo in the south. There is, however, one effective counter-strategy: take away its leaves. This is, of course, exactly what is happening as land-hungry people cut into rain forests in Panama and throughout the hemisphere. So far, most sloth populations remain secure in their environment. But along the Brazilian coast, less than 3 percent of the original rain forest has survived human habitation, and one sloth species, *Bradypus torquatus,* is endangered as a result.

Scientists know almost nothing about *B. torquatus,* the maned sloth. It possesses the fixed grin of other three-toed sloths, and a distinctive cowl of dark fur over its shoulders and down its back, which makes it look rather like an idiot prince. At the Poco das Antas Biological Reserve in Rio de Janeiro state, researcher Laurenz Pinder has begun radio-tracking maned sloths as part of a World Wildlife Fund project. But in the entire five-thousand-hectare reserve, with trained naturalists constantly on the lookout, only about fifteen maned sloths are sighted each year. So far, Pinder has put radio collars on just six animals. Outside the reserve, in the isolated pockets of rain forest surviving on the coastal strip between São Paulo and Bahia, the species is apparently even less common. One of Pinder's objectives is to discourage people from taking the animals as pets or from shooting them in the course of indiscriminate hunting.

The question of the sloth's eventual survival or extinction brings us finally to the matter of its supposedly abysmal stupidity. The idea that sloths are stupid, rather than merely simple, got its start centuries ago, when writers began repeating a myth about the animal's dietary ineptitude. According to Buffon and others, the sloth would climb a tree and gradually strip it of every available leaf. Then, being incapable of climbing back down, it would simply drop to the floor of the rain forest and drag itself off to kill another tree. A second myth was that sloths would eat the leaves of only a single tree genus, the *Cecropia*. Without it, as Buffon suggested, sloths would cease to exist.

But in their study on Barro Colorado, Montgomery and Sunquist found wonderful elegance in the way sloths harvest the jungle foliage. Though they are solitary animals, they have evolved a neat social system. Sloths in fact eat the leaves of at least ninety species of trees and vines, and different individuals prefer different species. Each sloth has a range of eight or so favorite trees and moves among them via the treetop network of vines. Thus several sloths can live in the same territory, not competing with one another but making such efficient use of the trees that they inhibit other animal species from moving in on their ecological niche. According to Montgomery and Sunquist, a square kilometer of healthy rain forest may contain seven hundred sloths and only seventy howler monkeys, their most numerous mammalian rival. The monkeys get all the attention, making a noise like a cruise missile whistling past one's earlobe. But it is the sloths who proliferate. The tropical canopy is packed with them. Evolutionary theorists have even speculated that the sloth's long-standing dominance in its niche may be the reason the New World tropics produced so few primates.

Sloths pass their dietary preferences on to their offspring. The infant (which Montgomery described as "a round ball of fur, very soft, and with soulful eyes") spends its first six months or longer clinging to its mother's hair, learning to eat what she eats. In the process, by licking bits of leaves from around her mouth, it also picks up her mix of gut microorganisms, which are specific to her preferred list of edible leaves. (Three-toed sloths in captivity have starved to death on a full stomach because they were fed leaves for which they did not have the proper digestive microorganisms.) And when the mother turns her offspring out on its own, she gives it a part of her range to help it get through the first year, after which it must strike out into the world.

There is an admirable harmony in this way of life. By exploiting a variety of different trees, sloths avoid putting pressure on any single tree species. And in Montgomery's interpretation, they may actually cultivate their preferred trees. By burying their excreta at the base of the trees that feed them, they return about half of the nutritional value taken out in leaf eating.

This elegant system may be the reason sloths can get by on about one-tenth of the work load of other mammals their size. It is the reason sloths can spend their mornings dozing in the sun while the rest of the animal world wearies itself with the daily toil of getting and spending. Sloths have adapted perfectly to their environment. They have made themselves masters of digestion, champions of sleep, gurus of the pendulous, loafing life. They will survive in splendid indolence as long as humans do not destroy their habitat. And *we're* calling *them* stupid?

The Devilbird
of Nonsuch

It is a changeable, rainy morning in Bermuda. Under the racing clouds, David Wingate stands at the helm of his Boston whaler looking like Zeus with eyeglasses. He has a Grecian nose, a full beard, and hair that swirls up in gray wavelets. The resemblance is utterly at odds with his personality. Wingate is modest and possessed of ungodly patience; he does not hurl thunderbolts, even under provocation. And yet the resemblance is appropriate because Wingate's life also has some of the qualities of a myth.

He heads the boat out across Castle Harbour toward a comma-shaped hump of land, an island on the edge of the open sea. The island is formed, he says, of aeolianite—sand piled up by the wind and hardened into rock. In fourteen and a half acres, the island contains beaches, forests, a valley, two marshes, and one human habitation, where Wingate lives as caretaker. Called Nonsuch Island after a favorite palace of Queen Elizabeth I, the island is like no other place on earth. Wingate set out almost forty years ago to make Nonsuch unique, and he will be at it for at least another ten years. When he first came to work on the island, humans had reduced it to a "desert" populated largely by rats and goats living among the skeletal remains of the old forest. When he is finished, he will have pieced the island back together into something like the way it was before humanity ever laid eyes upon it. The island is the focus of his life. He is restoring it not simply for its own sake but as a sort of bridal bower awaiting the return of a bird, the cahow, believed to be extinct for 350 years. Wingate, now age sixty-three, is still not sure he will see the bird return during his lifetime.

By his own account, Wingate was one of those odd children who are called to nature almost from birth, and he loved it with an intensity that utterly bewildered his father, a civil servant, and his mother, who played golf. As a toddler on Bermuda, he once took a jar of spiders to bed with him, and his mother found him asleep, covered with webs. His parents gave him guidebooks and binoculars and stood back, and Wingate spent his childhood at a place called Spittal Pond, where he logged his first swallows, his first egret, his first sandpipers.

His most vivid memory—a sort of epiphany—is of visiting the pond during a 1948 hurricane: "The sea had cut through and turned the pond into a small bay. I wandered down to the water's edge with my dog, and then I saw this bloody tidal wave. I was scrambling up the rocks to get away and my dog was running and then the wave came. I was on an isolated hillock and the water was all around and I realized, 'Christ, I'm on an island,' and then the water receded." At about the same age, Wingate began to explore the rocky group of islands that includes Nonsuch. He was looking, among other things, for the cahow.

Wingate does not put it this way, but the cahow is roughly Bermuda's equivalent of the dodo, and in the annals of extinction, the story of Bermuda and Nonsuch closely resembles that of other oceanic islands. Beginning around 1600, European explorers unwittingly ruined most of the remote islands of the world.

They introduced goats, pigs, cats, dogs, and rats, which combined with human appetite to wipe out a menagerie of benign creatures that had flourished in peaceful isolation. Writing about Tahiti, Alan Moorehead once summed up the European arrival in a phrase that has just the right suggestion of deadly speed. It became the title of his book: *The Fatal Impact*. Before Daniel Defoe wrote *Robinson Crusoe,* the island on which he modeled Crusoe's idyll was already overrun by rats. Even then, the quintessentially unspoiled island had become a myth.

In his copies of the early histories of Bermuda, Wingate has lightly marked the sections that describe the impact of humanity on the cahows. There were perhaps a million of the birds before the first settlers arrived in 1612. This pigeon-sized petrel with a thirty-five-inch wingspan spent much of its life hundreds of miles from land, airborne at sea, usually out in the Gulf Stream, for months at a time. Yet its nest, to which it returned each year after reaching maturity, was a long burrow under the roots of the cedar forests of Nonsuch and a few other islands. During the nesting season, from January to June, the birds emerged at night to fill the air with "a strange hollow and harsh howling." Spanish sailors nervously joked that they were devils.

In fact, the cahows proved a godsend to the earliest settlers. The first minister to the British colony wrote of "silly wilde birds coming so tame into my cabbin" that it was "as though they did bemoan us take, kill, roast, and eat them," which is what the settlers did, especially during famines. The cahow population thus made a quick march into oblivion, and the last recorded sighting of a cahow was in 1620.

The idea that the cahows may someday return may seem quixotic, and Wingate's efforts to restore Nonsuch for that event have at times been marked by unsettling tribulation, outrageous wrongs. But one gets the sense that Wingate feels something like chagrin about only one thing: that he did not single-handedly rediscover the cahow. Instead, in 1951, at the age of fifteen,

Cahow

Wingate was merely invited along when two venerable ornithologists—Robert Cushman Murphy and Louis Mowbray—undertook the search, and he was only an assistant when they found their first cahow chick on a rocky islet near Nonsuch. Someone else got to exclaim, "By gad, the cahow!"

But Wingate got the job of doing something about it, and what he did, after studying ornithology at Cornell University, was to begin rebuilding Nonsuch with funding from the government. Tending to the eighteen surviving cahow couples was not really a full-time job. Wingate had time to think well into the future, to a time when his fostering care might yield a thousand couples that would eventually overflow their islets and spread back to their old nesting grounds on Nonsuch. The ideal thing would be not merely to preserve Nonsuch for them, but to put back the same trees and shrubs and animals—down to the whelks and even the hermit crabs living in the shells of the whelks—that were there when the cahows flourished: He wanted to reverse the fatal impact, to rebuild the stage for evolution.

As he tucks a small cedar into the island's shallow soil, Wingate mutters to it: "Now, you son of a gun, grow. Don't die on me." Most of them die anyway because of a scale insect pest that in the 1950s denuded Bermuda of its most characteristic tree. Wingate, who had counted them, says that all but one of the two-thousand-odd cedars on Nonsuch died. Louis Mowbray persuaded the government to declare the island a nature reserve anyway. (It had formerly been a yellow-fever quarantine station and then a reform school.) Shortly afterward, Wingate arrived on Nonsuch to find that a U.S. Navy survey party had stormed the beaches and sawed down the only remaining live cedar because it got in the way of a guy line for a communications system antenna.

Today, more than three decades later, Wingate's island is a semitropical semiparadise. He stands in a verdant lane under a twenty-foot-high cedar and palmetto canopy and remarks, "Everything you see that's green, I planted." The navy incident caused Wingate to move to the island full-time with his wife and baby. Nonsuch became their whole world. "Rather than go ashore, I shot the goats and curried them. My friends thought we'd gone bush." Once, when a New Year's Day gale smashed their boat, they were marooned on the island. On the fourth day of the storm, they ran out of baby food, and Wingate was forced to swim fifteen hundred feet across to the mainland for help. "But those were the happiest years of my life," he says, "and that's when I got the idea for all this."

There were days when he planted two hundred seedlings by himself, inspecting each one to ensure that it did not carry stowaways, such as the dime-sized whistling frogs that were accidentally introduced into Bermuda in the 1880s. For five years, he got no results and doubted whether he ever would. "It was all very long-term," he says. "Not a cosmic time frame or even a geologic time frame. But you think in a much longer time frame than, say, a politician does."

Wingate's way of thinking was also unusual in that the native plants and some of the animals he was bringing back were, to use his own rather harsh word, dreary. Bermuda itself offers all the splendid flora of a tropical paradise—huge hibiscus flowers, oleander hedges, royal poinciana trees that become a mass of red blooms in the summer. But Wingate points out that these are all exotics imported from elsewhere in the world—all, in other words, false. He wanted only native species on Nonsuch. The flora on mainland Bermuda is 95 percent foreign; he aimed for Nonsuch to be at least 95 percent native (and says he has achieved that goal).

Wingate is not dogmatic enough to argue that dreary is necessarily better. "There is no question Bermuda is a more beautiful, diverse place," he says. "It was a very simple ecosystem before—it almost cried out for a little variety. There were no colorful flowering plants in the original forest; there may not have been bees to pollinate them." But he adds, "Bermuda the way humanity found it was God's creation or the result of a gradual evolutionary process. Primitive Bermuda was stable for tens of thousands of years; any other Bermuda I might have re-created was a very ephemeral thing." Primitive Bermuda was also, of course, the Bermuda that fostered the cahow and other fauna, and this thought makes Wingate almost jaunty on the subject of dreariness: "If it's dreary, so what? The achievement will be the extent to which I can bring back that dreary state. The important thing is that it's unique. We're putting enough effort into beautiful botanical gardens in Bermuda."

At night these days on Nonsuch, the only light is the glow from the U.S. military air base across the inlet. The only sound is a rustling along the sides of the path, which a flashlight reveals to be the scuttling of land crabs. The crabs start Wingate talking about his early failures, of which he readily admits there were many.

One of the islets off Nonsuch had been a nesting site for a bird called the Audubon's shearwater. "But I was too late to save them," he recalls. "We were down to one pair when I first came here, and you could hear their plaintive call at night. Then one died and the partner came back alone for ten years. I went out and recorded its call and then played it back, and the poor thing came out and copulated with the tape recorder. It was tragic." The species is now gone from Bermuda.

In another phase of the Nonsuch restoration, Wingate planted scores of trees that the nursery told him were endemic palmettos. They turned out, much later, to be of the wrong species, an import. "It was soul-shattering to have to destroy a forest and start again," he says. "We lost ten years of growth."

When their children reached school age, the Wingates had to give up their life on the island. But they returned from the mainland during summers and school vacations. At about this time, his wife died of burns from a kerosene-lantern accident on Nonsuch. "The whole thing seemed to lose meaning for a while." Wingate arranged to have her buried on the island in an old cemetery formerly used for victims of yellow fever. He planted a native tree, a yellowwood, beside her grave.

Then things began to turn. The disaster with the land crabs suggests the change that took place. "My mistake on this one was my impatience to see things grow," he says, thinking back to his first years on Nonsuch. "I thought that if I used fertilizer it would help the trees. I really poured the stuff on and, man, how they grew."

But so did the population of land crabs. Wingate soon found that they were eating the organic fertilizer and flourishing on it. Much of the island became spongy and hazardous with land-crab burrows. At night, the paths became a swarming Times Square of the land-crab world. It might have been a neat little lesson not to play God with nature, even when you're merely trying to put things back the way God or nature intended, except that the disaster eventually worked to Wingate's benefit.

In the back of his mind, Wingate had been planning to reintroduce the yellow-crowned night heron. For this heron, crabs are culinary heaven, and on Nonsuch the herons had evolved shorter legs and stouter bills, the better to pursue their prey in the undergrowth. But the local race of herons had been exterminated by early settlers. Wingate decided to bring in cousins—a different race of the same species—from Florida, where crayfish and blue crabs are standard fare. "The best thing we could do," he says, "is to introduce the nearest relative and hope to get evolution started again." To feed the nestlings and accustom them to their new environment, Wingate had to collect two hundred crabs every night. Under the circumstances, it wasn't difficult.

There are now more than a hundred herons living on Nonsuch and mainland Bermuda, the land-crab population is under control, and the pathways on both islands are solid underfoot. It may be that Wingate talks so freely about his early disasters at least partly because so many things have lately begun to work out so well. Today, he says, the Nonsuch restoration is "a string of unexpected successes." He admits to a "godlike feeling" at the thought that certain plants and animals are at home on Nonsuch because of him. "The really exciting thing," he says, "is to create a whole ecosystem. You begin to see synergistic effects. Two things by themselves are great, but two things together have a mutually supportive effect that is better. Things begin to fall into place." For instance, the reduction in the land-crab population, combined with the removal of introduced rats, has enabled a unique understory flora to recover, including the extremely rare endemic Bermuda sedge. Wingate calls this sedge "the botanical equivalent of the cahow," and he asks, "Could any biologist have predicted that the survival of this sedge depended on a crab-eating heron?"

Wingate, who is now conservation officer for all of Bermuda, is sufficiently encouraged to predict that within his lifetime there will be a mature, native forest on Nonsuch. But will the cahow return to live in it? "Ten years ago from the porch of the house," he says, "I heard cahows calling in the bay. I never thought I'd hear that in my lifetime." Now they fly over his house on the island and call hauntingly to one another, and to him.

In the 1950s, Wingate watched his cahows produce a maximum of four chicks a year, then eight in the 1960s, and seventeen in the 1970s. A cahow takes eight years to mature, and Wingate was beginning to see the offspring of the chicks he had fostered in the 1960s. In the 1980s, the cahows regularly produced more than twenty chicks a year, and now the fledgling crop is creeping close to thirty.

He has been a foster parent to ten of the birds. He has taken orphans and reared them in bunkerlike burrows he has constructed on Nonsuch, bringing the birds out each day through a special hatch for feeding. At the end of the nesting period, the parents abandon their young; then for seven nights in a row, the nestling emerges from the burrow to stretch its wings in anticipation of flight. Finally, it climbs to the highest available point, takes off, and heads straight out to sea. Wingate has always tried to be present for this occasion. He brings out a mattress and lies down near the mouth of the burrow to wait and watch. Once one of his orphans used him for its takeoff point. "He pulled himself up, working his bill along with a sort of vibratory nibbling," Wingate recalls. "He gave me a good hair pull and an ear nibble and he shat, as they always do before they take off. Then he went straight up like a helicopter. It was very satisfying." But even the orphans do not yet return to Nonsuch to breed. They prefer to nest near other cahows, and the population—now up to fifty-three nesting pairs—will spill over into Wingate's forest only when the islets around Nonsuch have become too crowded. "Psychologically, it's a very big step for them to colonize a new island."

But Wingate has a pretty exact idea of when that will happen. He thinks about it some nights when he carries his mattress out to a point of the island to see, or perhaps feel, the cahows winging overhead. "The nature of exponential growth is that it seems agonizingly slow at first," he says. "But suddenly, you begin to notice the acceleration. Eventually, you're going to have a thousand pairs of cahows, and not long after that it will be two thousand."

Wingate keeps his night watch on the island promontory at least in part to get a sense of what Nonsuch will be like then. He talks about the sound of the cahows calling to each other weirdly out on the bay. Sometimes, he says, they shoot past two or three feet over his head. "It's beautiful flight. They make sweeping figure eights . . . all without any visible beating of the wings."

He expects that one of the cahows' islets will spill over first to the point where he keeps his night watch. Then the population will gradually spread another hundred yards, past the spot where the last of the old cedars was cut down. It will expand around the hook of the island, past a salt marsh he has created. Finally, around 2050, the cahows will be breeding "by the thousands" in the center of the island, on the floor of the palmetto and cedar forest he planted.

Of course, Wingate will be dead then. With continued local and international interest, someone else will have taken his place as caretaker. He hopes to be buried on Nonsuch in the old cemetery. The island has always been a part of him; then he will be part of the island. And he dreams the cahows will breed on his grave.

The Cave of
the Bats

On Route 7 a few miles north of Manchester, Vermont, a tree-covered, thirty-two-hundred-foot knob of limestone and marble, once a source of New England gravestones and the raw material for national monuments, divides the villages of Dorset and East Dorset. At various times and depending which side you lived on, it has been known as Elk Mountain, Green Peak, Dorset Mountain, and Mount Aeolus.

The last name, a high-minded but misguided choice, seems to have stuck, and recently became official. It dates back to the fall of 1860, when an Amherst College geology class climbed to an observation point high on the mountain's eastern flank and lodged its protest against the "low or vulgar" tendencies of New England geographic names like Hogback and Potato Hill: "What an improvement upon Rattlesnake Mountain would it be to call it Mount *Crotalus* (the Latin name for this snake)," their professor wrote afterward. "What a mighty gain, I may add, would it be, to substitute Mount *Leo* for Camel's Rump." To his class in Dorset that day, the professor made a brief peroration likening a cave on the mountain to the one in which the ancient god Aeolus restrained the winds. Then he shattered a bottle of water with his geological hammer and christened the mountain Aeolus, to be pronounced *ay*-a-lus or *ee*-a-lus, possibly depending on which Dorset you lived in.

One hesitates to add to the nomenclatural confusion, but the plain, vulgar, Anglo-Saxon truth is that he should've called it Bat Mountain. "The Cave of the Winds," which harbors no winds to speak of, should have entered history as the Bat Cave, the name it is still commonly given on both sides of the mountain.

Located twenty-four hundred feet up, at the foot of a mossy stone outcrop, the cave draws bats by the thousands from as far away as Cape Cod, and they are the mountain's most distinctive feature. The cave serves them as a sort of ultrasonic discotheque in late summer, when the sexual debauchery matches that of certain Manhattan nightclubs; in winter, it becomes their hibernaculum, quieter than a monastery, its ceiling matted and ribboned with torpid bats. Nobody honestly knows how many bats use the cave, long-term or as transients, but estimates have ranged from two thousand to more than three hundred thousand. The cave on Mount Aeolus, while minor relative to some of the teeming bat caves of the Southwest, is probably the most important bat-swarming and -hibernating site in good, granite, cave-poor New England. In this century, it has also played a vital part in helping explain how bats live and why they should matter to the humans who mistakenly dread and despise them.

The cave now belongs to The Nature Conservancy, part of a 150-acre property donated by a mining company, National Gypsum, in 1983. Mark Des Meules, a Conservancy scientist with prominent brown eyes and the rumpled, sleepy look of the perennially overworked, was its caretaker when I visited one February for a midwinter bat census. Des Meules explained that the cave mouth is gated off now in winter, because of a history of sporadic vandalism. In China, bats are symbols of good luck, but Westerners have long regarded them with almost pathological loathing. This attitude has been compounded since the 1950s by an exaggerated connection to rabies, a disease humans are far more likely to acquire from dogs, cats, skunks, or foxes than from bats. Over the years, troubled souls have felt justified in entering the bat cave to slash hibernating bats with knives, incinerate them with torches, and deafen them (as well as themselves) with shotgun blasts.

A bat's life is perilous even without such assaults, Des Meules said. In temperate climates, bats live entirely on insects, and every evening of the summer, a bat must catch and kill roughly half its own body weight in insects. According to one estimate, 50,000 bats live in a typical 100-square-mile section of New England, and they consume 13 tons of insects over the course of a summer. But with the coming of cold weather, the insects disappear and the bats hang themselves up like out-of-season garments for the next five months. Their body temperature plunges and their heartbeat slows from thirteen hundred times a minute in flight to as little as eight times a minute in deep hibernation. Until April, when the bugs reappear, they must sustain themselves with stored fat. Merely waking them before then can burn ten to thirty days' worth, enough to keep the bats from surviving the winter. To protect them, the Conservancy designed a gate of angle irons spaced so that bats, but not people, can get through. Except for infrequent research visits, it stays closed from September through May, and I can report that it works.

Having gathered from various remote parts of the region, through some of the worst driving weather of the winter, the bat census crew made the long, switchbacking trek up to the cave in snowmobile relays. The gate was held shut with a padlock on a tempered steel chain, and the man with the key fumbled with it briefly. Then the key—the only key within a two-hour drive—dropped and slid on the ice, slowly, horribly, as if drawn by the bats themselves, down through the gate and into the cave, out of reach. We attempted to squeeze through the bars, and briefly

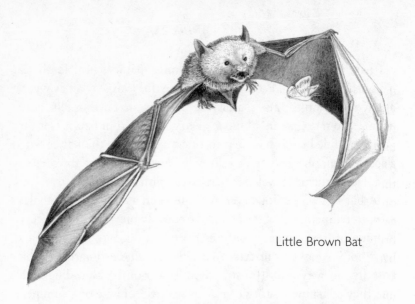

Little Brown Bat

mooted the possibility of gaining admittance with the help of small children or large explosives. But it was as if the long-suffering bats had put out one of those welcome mats that say "GO AWAY." Having little choice, that's what we did.

Bat researchers have been coming to Mount Aeolus since 1934, when a local naturalist introduced the cave to a Harvard instructor named Hal Hitchcock and one of his undergraduate students, Donald Griffin. Scientists then knew almost nothing about the biology of bats, which represent roughly a quarter of all mammal species. The two researchers were to shed substantial new light on this nocturnal and elusive subject.

Griffin, who has tarnished silver hair and wears rectangular, horn-rimmed bifocals, grew up collecting animals when he was supposed to be in school. Having started out with birds, he began to band bats as a teenager in Barnstable, on Cape Cod, where large summer colonies of mothers and their young roost in cottages near the water. At Harvard, Griffin wanted to branch out into caves, and Hitchcock went along for the ride.

Researchers then didn't believe it was harmful to disturb hibernating bats, and Hitchcock recalled that on their spring and fall visits they plucked the torpid bats off the cave ceiling "like grapes" to tag them with small metal clips. While Griffin was still an undergraduate, they picked a bat off a wall at Mount Aeolus and found one of his tags from Cape Cod—the first documented proof that bats could migrate long distances. Another Griffin tag, placed on a Cape Cod bat in 1937, turned up at Mount Aeolus in 1960, establishing a longevity record of twenty-four years for little brown bats, the cave's most abundant species. Hitchcock later beat this when one of his bats was recovered after thirty years, in 1977. These records are more than mere curiosities. They refute one of the major human misconceptions about bats, whose small size and tendency to cluster in large groups suggest the brief, prolific lifestyle of the thoroughly expendable mouse. But bats are more complicated than mice, to which they are related more distantly than to human beings. As one biologist puts it, "You're talking about little bitty creatures with the lifespan of a black bear."

With other researchers, Hitchcock went on to establish that the bats from Mount Aeolus migrated in summer, fanning out over a broad area from Hartford, Connecticut, to Manchester, New Hampshire. In one case, a bat banded at the cave turned up three nights later in Brookfield, Massachusetts, a trip of eighty miles for an animal that had only recently emerged from hibernation. Many of the bats also made long reconnaissance flights back to the cave in midsummer, perhaps to teach their young how to find the hibernaculum.

Griffin meanwhile turned away from migration and Mount Aeolus to the question of how bats fly in the dark without smacking into obstacles. An Italian researcher named Abbé Lazzaro Spallanzani first demonstrated in 1793 that bats do not depend on their eyes in flight. Spallanzani actually blinded several bats and observed them on the wing. Then he recaptured them several days later, dissected them, and found their bellies crammed full of insects, demonstrating that they could find and catch small prey despite the handicap. He came to believe that bats "see" with their ears. But later scientists ignored or ridiculed Spallanzani's work.

The leading theory in the nineteenth century was that bats navigated by somehow "feeling" the reflected sound waves from their wingbeats. In 1920, an English physiologist hypothesized that high-frequency sound might be involved, but he never tested the idea.

With a cage full of bats, Griffin presented himself one day in 1938 to a prominent Harvard inventor and professor of physics named G. W. Pierce, who had lately developed a device for detecting sounds beyond the range of human hearing. Pierce was intrigued, and when they turned the device on the cage, bat communication—"a medley of raucous noises," clicking and sputtering—became audible for the first time to human ears.

To test the idea that the bats might be navigating with the help of these "supersonic" sounds, as they were then known, the two men let some bats free in a room. Pierce's device picked up occasional bursts of sound, but to their disappointment, it was mostly silent. Griffin eventually figured out that the microphone they were using detected sound only when held directly in front of a bat's mouth, a serious impediment: A bat dipping and veering around a room will not hug a mike in the manner of Bing Crosby crooning "Going My Way." With a fellow student named Robert Galambos, Griffin then devised a series of experiments to determine how well bats flew around obstacles in a room when their ears were plugged, or their eyes blindfolded, or their mouths muzzled. The result was the first description of animal sonar, or echolocation—the ability to produce short bursts of high-frequency sound and read subtle nuances in the resulting echo to form a detailed picture of the world. Griffin later established that bats use echolocation not just to dodge obstacles, but to find and capture their huge daily intake of bugs.

Scientists now know that a bat echolocates by producing anywhere from five to five hundred chirps per second, with the rate climbing rapidly as it closes in on a target. It can employ echolocation at relatively long range; if you flip a penny in the end zone of a football field, a bat at the five-yard line can spot it in the dark. Bats can also detect other bats on a hunting ground at a distance. But the bat's echolocation works best up close, in tandem with the extraordinary controlled flexibility of its wings, enabling it to roll and slip and spiral down in hot pursuit of the most agile insect. (The bat scoops up its prey with the wing membrane between its hind legs, then plucks it up with its teeth, while continuing to fly.) Writing in an earlier decade, Griffin subtitled a section on bat foraging "Tally ho! June Bug at Two O'Clock."

But our knowledge of echolocation has advanced considerably since then, and researchers now speak of the bat "giving away information about itself" with every sound pulse, allowing insects to take evasive action. According to one theory, tiger moths actually produce a high-pitched sound to jam the bat's sonar, meanwhile folding their wings and diving for cover. (Other researchers argue that the moth is merely announcing that it tastes bad, or is attempting to startle the bat.) The advantage still seems to be with the echolocating bat: In one of his laboratory studies, Griffin found that a single little brown bat caught 175 mosquitoes in fifteen minutes, or one every six seconds. What is going on out there is an infinitely more sophisticated version of *Top Gun,* conducted by an animal with a brain the size of a pencil eraser. If you subscribe to the idea that mosquitoes are the enemy, this means the long-maligned bat in truth deserves the sort of adulation otherwise reserved for the likes of Tom Cruise.

Mark Des Meules had rescheduled the bat census for the end of February, and we met again in East Dorset on the heels of another snowstorm. Arnold Elithorpe of Vermont Fish and Wildlife got out his snowmobile, strapped on a magenta sparkle Arctic Cat helmet, and gunned the engine, announcing his intention to "Give her the sugar and hope for the best." Then we headed up Mount Aeolus, with multiple keys to the cave gate in hand.

This time, we got the lock off, but the gate was frozen solidly shut. The cave exhaled, sending up smug clouds of vapor from somewhere in its warm interior, and the rocky outcrop was feathered with rime ice. Des Meules attacked the gate with rocks, boots, and Elithorpe's penknife, before it finally yielded, and one by one, gingerly, we eased our way inside.

The cave entrance, dubbed "Guano Hall," slopes down steeply and was coated with ice and littered with angular rocks for about eighty feet. It was empty of bats now because of the cold, but in summer, all the action takes place here. One August evening in 1961, Hal Hitchcock netted 663 bats at the entrance. "After dark if you stand at the mouth of the cave, they just seem to come out of the bushes in every direction, coming in," Hitchcock recollected, when I phoned him afterward. The cave still lived in his mind's eye, though in his old age he was blind and disabled by a heart condition. "They just spend one night there, apparently, and then leave. I was in there once when the air was literally full of them. It seemed to me that they were flying in a more rapid and agitated fashion, just zooming around in that great big open chamber, and passing in and out of all the little passages. It was quite a sight."

The agitation may have had something to do with what Brock Fenton, a researcher at York University in Toronto, describes as a "disco mating strategy." Bats, he has written, aren't "sterling examples of loyalty or domestic tranquility." In late summer, when they are in peak condition, "just coated with fat," both sexes come to swarming sites like Aeolus. The males cling to the walls and direct squeaks, squawks, and ultrasonic wolf whistles at the females, who cruise up and down the hallway, stopping to spend time with several different males along the way. Fenton's description of mating is not edifying (moral readers may therefore wish to skip to the relative safety of the next paragraph): "The male approached the female from behind, typically embraced her with his forearms, and grabbed her by his teeth at the scruff of the neck." As well as being a brute, the male is unusually well-endowed, because the wing membrane between the female's hind legs requires him to cover a greater distance for the sexes to meet. Fenton adds that other bats like to hang around to watch "and the antics of the spectators often result in the termination of the coupling." Then they move on to new partners.

All this must of course be a terrible disappointment to people who believe that the better sort of animals all form lifelong monogamous unions, or perhaps it will merely confirm their low opinion of bats. But Fenton suggests that "random and promiscuous" mating makes sense for bats because hibernation follows soon after, and the bats are then helpless to protect their investment in a carefully chosen mate. The demands of monogamy would merely waste precious energy.

With Des Meules leading the way, we picked our way down into the hibernaculum, using the ice stalagmites for footholds. Noise disturbs the bats, so we were whispering now and trying not to send chunks of ice or rock clattering down ahead of us. At the foot of the slope, Des Meules played his flashlight across an empty patch of ceiling. "There's usually a mat of them here," he said. But it had been the harshest December on record, and it was possible that the wedge of cold air coming down from the cave mouth had driven the bats farther back. He pointed to a dark spot in a patch of ice on the ceiling. "That guy actually looks dead. He's frozen in there. He chose the wrong spot."

We moved deeper, past the throat of the cave, and into the rooms beyond, where it was warm enough for water to seep steadily off the ceiling. "Here we go. Fourteen ML," Des Meules whispered, using shorthand for little brown bats, whose scientific name, *Myotis lucifugus,* means "mouse-eared" and "light fleeing," a fair name. Little browns are drab, mousy creatures, lacking the outsize ears and unusual facial ornaments of some other bats. When their mouths are open and their insect-gnashing teeth exposed, they look like the sort of particularly unpleasant lap dog that a Barbie doll would have in her spiteful and neglected old age. (Bats always have their mouths open in flight, to produce the endless chirping of echolocation, and this is one reason people who see them in photographs mistakenly believe they are ferocious.)

In hibernation, the bats looked very nearly dead. Some, off by themselves, were as emaciated as a leaf that has hung on into spring, withered and glistening. Others, covered with frosty beadlets of condensation, looked at first like something that was left in the freezer too long. Up close, I watched carefully and could just detect their slow breathing, like sleeping infants. The beadlets made some of them glitter like Christmas ornaments, white except for the tips of their ears and the black dots of their open eyes. The sharp claws of their tiny feet were spread out and hooked over the slightest nubbin in the rock face. A few also held on with a forearm angled out to one side. One of the census crew found a bat on the ground in a narrow side passage. Weasels or mice usually clean up such casualties, but this one was still alive. We picked it up and hung it back on the wall like an old coat.

Des Meules called off sixty-nine little brown bats, and one northern long-eared bat. (Altogether six bat species are known to use the cave.) He climbed up into a sort of overhead pothole, and ran his flashlight across a mat of animals, packed twenty or thirty deep in places. A few of the bats were beginning to stir now. Some of them swayed visibly, as they shivered toward wakefulness, warming themselves up at the rate of about one degree a minute. Some shifted their forelimbs, making the slow, dreamy, ineffectual movements of a person trying to fend off a nightmare. They dreamed perhaps of oafish, brutal human beings bearing torches.

Up in the pothole, Des Meules located a cluster of seven Indiana bats, an endangered species, which has begun to return to the

Mount Aeolus cave since it has been protected by the gate. It was the best showing for the species in recent years. But a rough count of the total bat population was low, down from about twelve hundred a year earlier to fewer than five hundred now. Perhaps the other bats had gone to mines or caves elsewhere in western New England, or across the border to New York, or perhaps they had moved back into other parts of this cave where humans cannot reach. If the Aeolus cave has no winds, it does at least have a slight draft, indicating that there might be another opening, Des Meules said. Then, overhead, we heard the urgent, leathery flapping of a bat's wings in flight. It was time to go.

Left to themselves, the bats will wake up in April, possibly, according to Fenton, "under the nervous strain of a full bladder." The females will wake first, to eat and get a start on the business of bat motherhood.

Despite the orgiastic night in the disco six months earlier, the female is not yet pregnant, nor can she expect much help in this regard from the male, who may sleep for a solid month more and who is, in any case, useless for further romance. A bat's fat content, roughly 30 percent of its body weight when hibernation began, is down to 5 percent on waking; the newly awakened male has no energy, no *joie de vivre*. The bat's extraordinary adaptation for getting around this impasse is that, unlike any other mammal, she can store sperm. All through hibernation, she has somehow kept the seed of several males alive in her womb, without suffering any visible immune system havoc. Only in April, when food is available to support a pregnancy, does she finally ovulate. Some scientists theorize that she may actually have some mechanism for determining which sperm is most suitable for fertilization.

It occurred to me that a newly awakened female wouldn't have the energy for pregnancy. But Tom Kunz, a Boston University biologist who has studied bats at Aeolus and at maternity colonies in New Hampshire, explained that bat gestation is slow, up to sixty days, and undemanding in the early stages. If food becomes unavailable, the female may simply slip back into torpor for a few days, slowing the process down.

Even so, the marginal character of life as a flying mammal soon becomes evident. Birds lay eggs and spare themselves from having to fly around with all that extra weight. But a bat gives birth to a single offspring weighing a quarter of her normal body weight, or sometimes to twins. (To ease birth with the help of gravity, she may hang rightside up and catch her newborn in the wing membrane between her legs.) Her young may then latch on to a nipple and cling to her in flight for several days afterward. Birds can gather food and regurgitate it for their young back at the nest; bats can't. The mother must continue to eat for two, nursing her offspring until it is capable of taking flight to forage for its own food. Since a young bat starts to fly when it reaches 80 to 90 percent of its adult weight, this is the equivalent of nursing a teenager.

In his studies, Kunz has found that bat milk is richer than that of any other mammal except seals and sea lions, allowing the bat to wean her young in just three or four weeks. But during that time, she must eat her weight in insects every night. "Think of the selective pressures that have led to these adaptations," Kunz exclaimed.

I was thinking of what an unsuspected blessing all this was for Cape Cod, Boston's North Shore, the Berkshires, the cities in summertime. I thought of the dormant, vulnerable bats on the cave ceiling, and, as Kunz suggested, of all the subtle shifts in behavior and physiology with which evolution had equipped them for the work they would soon be doing in the night skies all over New England. At times in Central America, the bat has been the object of religious devotion. One can get carried away with this sort of thing; Mayan bat worshippers practiced human sacrifice, for example. But even in our more modest and secular era, it seemed to me that Mount Aeolus deserved to be treated as a holy place.

Good Scents

One May afternoon in Connecticut, three ten-year-old girls on a class camping trip accidentally took a left turn instead of a right, and became missing children. The girls' parents spent a windy, wet night at a command post in the park, listening to searchers radio back about their progress through the dense, rocky terrain and imagining what they might find. It was easily the worst night of their lives. Ambulances were waiting nearby at dawn, and a hundred or more fresh volunteers were ready to renew the search. Among this group, the least impressive was an ancient bloodhound, red-eyed and loose-fleshed, who was called Clem. His handler, a state trooper named Andrew Rebmann, harnessed Clem and walked him to the last place the girls had been seen. The hound then sniffed at an article of clothing that one of the girls' parents had brought from home. "Find 'em," Rebmann said.

Clem worked silently, pulling hard on the lead when the scent was strong, hesitating and casting from side to side when it faded. After three miles, he dropped his nose on a purple barrette. A mile

Bloodhound in Full Slobber

farther down the trail, sniffing the air now instead of the ground, the hound came to the edge of a swamp. Searchers had stopped short of this swamp before, not believing that three ten-year-olds would have gone into it. This time, encouraged by Clem, they shouted, and from across the swamp they got a reply. Rebmann followed the hound through waist-deep water and, on the other side, on high ground, found the girls safe. They had gotten there by jumping from log to log, like Liza crossing the ice floes in *Uncle Tom's Cabin*. Rescuers carried them out again on their shoulders to greet their parents, medical attendants, and finally the press. Back home again, the three girls did the right thing: They sent Clem a bone.

In the annals of the breed, it was an unextraordinary case. The bloodhound, a large, doleful creature bred specifically to recognize people by scent alone, has saved countless missing children, and adults as well, often where large numbers of human searchers have failed.

The bloodhound isn't the most intelligent dog in the world, nor the easiest to train. The German shepherd is better at sniffing out certain kinds of objects—bombs, for example. The black Labrador retriever excels at searches that call for sweeps across a given area of land. But when it comes to following the specific trail of a particular person, the bloodhound has no equal. "It'll work and work, in any kind of weather, until it can't stand up anymore," says a bloodhound handler with the California Rescue Dog Association. "A bloodhound just won't quit on a trail."

Over the centuries, bloodhounds have chased one fugitive king (Robert I of Scotland), and been prized by two queens (Elizabeth I and Victoria). They have trailed sheep stealers, cattle rustlers, arsonists, murderers, and rapists. Bloodhounds have also helped recapture escaped convicts, though their real-life quarry has seldom been as sympathetic as some of the characters they have hounded in the movies—Tony Curtis and Sidney Poitier in *The Defiant Ones,* for example, or Paul Newman in *Cool Hand Luke.* When James Earl Ray, the convicted killer of Martin Luther King, Jr., escaped from Brushy Mountain State Prison in Tennessee in 1977, bloodhounds tracked him down.

Even so, the bloodhound is no popular hero. The problems begin with the bloodhound's physical appearance, which does not constitute heroic good looks. A good bloodhound has more dewlap than a Renaissance cardinal, and enough loose flesh at the back of the neck to form a clerical cowl. Drooping skin and long leathery ears are among the chief characteristics of the breed.

Bloodhound owners say that the ears sweep up the scent as the hound moves its head from side to side, and that the loose flesh, which falls forward when the dog has its nose "on the ground" tends to "cup" the scent. (Olfactory biologists generally agree that the hound's relentless cupped-nose-to-ground sniffing is an advantage. But when I raised the ear-sweeping theory, one scientist merely rolled his eyes and said, "Come, come.") Enthusiasts tend to have an endearing faith in the survival value of almost any bloodhound trait that happens to have been favored by human breeding. One owner actually argued that the bloodhound's loose skin is an adaptation to protect the eyes in collisions. His hound was so intent on the trail, he said, that he came to recognize the make of parked cars solely by the sound their fenders made when his bloodhound bumped into them.

The general droopiness extends to the lower eyelids, which, in some but not all bloodhounds, hang down, revealing wet, red underlids. These eyes belong to an opium addict, ravaged and yet sublimely calm. In extreme instances, the skin hangs down so far that the eyes themselves aren't visible. The raw, pink underlids look almost as if the eyes have been plucked out, contributing to the breed's ghoulish image.

The final—and some say ultimately damning—element of the bloodhound's appearance is called slobber by fanciers and froth by those who mistakenly believe the bloodhound is a cutthroat monster out of hell. Some fanciers argue that the excess of saliva is also functional: As the hound snuffs steamily about, it tends to vaporize dried-up particles of scent on the trail, making them more palpable to the nose. Others say that this is nonsense (most scent molecules are not water-soluble, but lipid-soluble), and that you too would slobber if your upper lips were flapping around several inches below their lower counterparts. In any case, the slobber is decorative for a bloodhound. It hangs down in ropes from the upper lips, or flews, and sometimes gets festooned between the lip and the end of either ear. When the bloodhound shakes its head, the loose flews snap slobber in all directions. It can thus become a decorative detail for the bloodhound owner, too, who may discover it midway through an important business meeting lying like an epaulet across the shoulder of his best suit.

Bloodhound owners, who otherwise share the normal human abhorrence for the word "sag," prize all this loose flesh. They like to explain that a bloodhound's flews hang down far enough to look right only if the hound can hold a beer can in his mouth undetected. (This is the sort of thing bloodhounds do for sport, with a look of droll amusement.) Owners and hounds together drive around in cars with license plates that say "WOEFUL" and even "DROOPY." Obviously, all this is a specialized taste, which may account for the relative scarcity of bloodhounds. According to the American Kennel Club, the bloodhound is sixty-fourth most-popular on the list of 145 registerable breeds.

The bloodhound's unfortunate name also contributes to its image problem. Fanciers of the breed spend a good deal of time patiently explaining that the creature just now setting its enormous forepaws on an honored guest's shoulders has little or no interest in actual blood. "Blood" got into the name, they say, because these were among the first purebreds, the first canine bluebloods. Modern bloodhounds descend from the hunting packs kept by monks and noblemen in the Middle Ages. The most famous of these packs was a strain of black hounds established in the seventh and early eighth centuries A.D. by a Belgian monk, François Hubert, who eventually became the patron saint of hunters. These were "blooded hounds," or bloodhounds, supposedly because they were bred from known bloodlines.

James Thurber, whose cartoons regularly featured bloodhounds of a benign and likable disposition, did not find this explanation wholly satisfying. "My own theory," he wrote, "is that the 'blood' got into the name because of the ancient English superstition that giants and other monsters, including the hound with the gothic head and the miraculously acute nose, could smell the blood of their prey. The giant that roared, 'I smell the blood of an Englishman!' had the obscene legendary power, in my opinion, to smell blood through clothing and flesh. . . . It seems to me . . . that legend and lore are more likely than early breeders and fanciers to have given the bloodhound its name."

A final possibility—one all bloodhound owners disdain—is that the name originated in a considerably more direct connection to blood. Some writers have theorized that bloodhounds were originally used to track wounded game by the drops of blood on the trail. Others have suggested that the bloodhound got its name because it tore its prey to gory bits. The name "hound" itself means "one who seizes," and in 1578, an English bishop wrote enthusiastically that bloodhounds had "such ferocity of nature" that when they catch up with villains "they tear them to pieces." An eighteenth-century poet rhapsodized on the bloodhound's "superior skill / To scent, to view, to turn and boldly kill."

This idea intrigued me, particularly when I came across the news in *Horse and Hound*, England's weekly hunt journal, that at least five bloodhound packs in that country now hunt humans. I could imagine the foxes getting quite chuffed with the idea, as they say over there. I could see them on the sidelines shouting encouragement and advice to the field: "Fine day for hunting, no? Got a glimpse of your quarry just now. Big, strong redhead in a Gore-Tex jogging suit. Went that way."

The puzzling thing was that the proposal would have caught on among the English people, in a modest sort of way. For a cap fee of about $50, according to *Horse and Hound*, an outsider could join the Windsor Forest Hunt on a Saturday when the weather was fair to hunt down three upstanding citizens of the Berkshire, Oxfordshire, Hampshire, or Surrey countryside. One might have thought this sort of thing went out with the Magna Carta.

So I caught the next plane for England, and soon found myself in the bucolic hills, sandwiched between the flight path out of

Heathrow Airport and the roaring traffic on the M4 motorway, said to be the busiest commuter road in Europe.

Hunting with bloodhounds has prospered, an official of the Windsor Forest Hunt was shouting at me through the din, because it is a more practical approach to the modern world. He was talking about overpopulation and all that, at least indirectly. The problem with traditional foxhounds is that no one can control where they go because no one controls the fox they are pursuing. This becomes awkward when the fox leads hounds and horses through winter crops, livestock, or a new suburb teeming with antihunt activists, or when, in the blind delirium of a hot scent, the hounds launch themselves onto a national motorway. Since the 1950s, fox hunters have been abandoning territory within five miles of motorways and in developing areas. Anyone who has traveled in England in recent years will know that this means almost the entire country.

Bloodhound hunts like the Windsor pack, which was founded in 1971, have moved into these openings. They have been able to do so because a human runner, given a twenty-minute head start, can lead a bloodhound pack on a carefully mapped-out two- or three-mile route, skirting all such modern hazards. Moreover, the Windsor Forest spokesman told me, "We don't do anything ghastly like killing a fox in the middle of the market."

The master of the hunt was a handsome, straight-backed Irishwoman named Ruth Coyne, who presided from horseback with a rather formidable mix of humor and indignation, crying out at hounds and humans alike, "Come *on!* Stop poncing around. Move on!" She was a great admirer of her hounds, which she knew by their names and personalities and ambitions for higher status in the pack. On the trail, she said, the bitches do all the work, particularly a girl named Delilah and two matronly figures known as "the fat old ladies." The males hang back and "bumble around."

The bloodhounds' first victim that day was a twenty-six-year-old salesman from Reading, dressed in powder blue sweatpants.

He was an agreeable fellow, a member of the "Banana Leisure" jogging club. Coyne leaned down from her horse to give him his final instructions: "You see that fence there? Go through it and go straight on. Get *on* now."

"O.K., right-o," the runner replied. He trotted off, leaving a sweaty T-shirt tied to a stake in the middle of the starting area, as the scent article for the hounds.

"He's got no smell," Coyne muttered dismally, when he was out of earshot.

The whipper-in had the pack of hounds gathered together at a safe distance, and now Coyne went up to greet them. "Oh, we've got the girlie, whirlie, pearlies!" she exulted. "The big fat ladies and the little thin ladies!" The hounds danced around her worshipfully.

"Bloodhounds hunt," Coyne called down to me, "because they like you. They want to please. They have to be allowed to have their fun, to play. They'll tell you to bugger off if you use the whip on them. Whereas the foxhound is tough as old boots."

As we waited for the runner to make the most of his head start, the chairman and founder of the hunt, a florid, overweight fellow named Major William Stringer, recollected his first efforts with bloodhound hunting as a British Army officer stationed in West Germany in the 1960s. "We used soldiers as runners," he said. "Volunteers. With a slight bit of pressure." He recalled in particular a Private Smith who volunteered on a day when a general from the German army joined the field. The general, said Stringer, "thought it was great hunting a human, absolutely perfect."

The unfortunate Private Smith apparently felt otherwise. But he set off, and after a reasonable interval, the hounds went after him, with the field of hunters behind. As the thunderous bellowing of the pack drew closer, poor Smith became increasingly terrified. "As we got on," Stringer said, "I suddenly noticed that the fences were getting higher, we were going over wires and all, and I thought, 'Where is this man Smith going?' And then I thought we were on to deer. Some of the horses were falling out, we were going so fast. I started looking for gaps and gates to go through. The general said, 'Come on, Bill, go on, give us a lead,' and I was thinking, 'My God, when I catch this bloody man Smith I'll give him hell.' Finally, I came up and there he was, hiding behind a tree. I was just about to put him in the brig, when the general came up and said, 'Nobody in the world can hunt as well as we did today.' So of course I said, 'Well done, Smith.' "

As Major Stringer finished his story, Ruth Coyne announced that time was up for the Banana Leisure runner. The whip led the hounds toward the starting point, and they were off in an instant, with hardly an olfactory glance at the scent article, their baying inaudible above the roar of the motorway. "Going like the clappers," Coyne declared with satisfaction, and then she galloped off, with her field of twenty riders in pursuit.

Major Stringer, who no longer rides, and various other friends and family followed the hunt by car, sometimes circling ahead of the pack. The frantic hounds ran directly where the runner had gone, or on a parallel course a few feet away where the wind had caused his scent trail to drift. Delilah and the fat old ladies scrambled under blackthorn hedges, burst through loose stone walls, leaped over obstacles with their loose flesh dancing around them like schoolgirls' skirts. As the scent got warmer, they raced faster and bellowed more wildly. Followers of the hunt describe this as "beautiful music." But one of the runners told me it was an "eerie, a most unearthly sound."

During the slower interludes, the talk among the hunt followers turned to the Labour Party's plans to outlaw foxhunting. According to the newspapers, two hundred protesters had showed up a few nights before at a hunt ball in Manchester, chanting "Rich bastards!" and spitting on women as they arrived in their evening gowns. Almost everyone in the Windsor Forest Hunt started out in foxhunting, and many still participate, so they tended, as a matter of form, to denounce animal rights protesters.

One of them worried that traditionalists might "think we're cashing in as a soft alternative to foxhunting."

Still, everyone was aware that squeamishness and unrest about blood sports are increasing, and no one asked indiscreet questions about anyone else's lily-livered sympathies. Queen Elizabeth herself decided years ago to abstain from the foxhunt. But hunting with bloodhounds has earned her approval. On one occasion, the Windsor Forest Hunt met by invitation in the royal park surrounding Windsor Castle and she joined several members of the royal family in the pursuit (the runners presumably being drafted from the ranks of the tabloid press).

Between the second and third run of the day, the hounds and riders were taking a rest in a field when we suddenly heard the snap-crackle-pop of machine gun fire. It developed that amateur war games were taking place in a woods nearby. Candy-cane-striped plastic police tape defined the battle area, where men and women dressed in gas masks and camouflage darted about, enjoying the outdoors and shooting one another. A judge in a neon green jumpsuit presided. The members of the bloodhound hunt surveyed this extraordinary scene in wonder.

Just then, a team of hunt saboteurs showed up at the end of the field to block the gate with their bodies in defense of animal rights. One of the more diplomatic members of the hunt gently explained that this pack consisted of bloodhounds; it wasn't actually hunting foxes, but . . . well . . . human beings. The hunt saboteurs contemplated this news, and after a moment, as if to say, "Well, that's alright, then," they allowed the gate to be swung open. The hunt proceeded to chase its final victim of the day.

The truth of course is that the bloodhounds did not kill anything that day or any other. The English people, while still known to dote on dogs, take a strong line against tearing the local citizenry to pieces. Indeed, the hunters would have a hard time getting their hounds to bite. Whatever fierceness bloodhounds may once have possessed has been bred out of them. Modern bloodhounds are unimaginably gentle, affectionate with children, and tolerant even of inquisitive journalists measuring the twenty-eight- to thirty-inch span of their ears. Bloodhound fanciers tend nowadays to deny that fierceness was ever a bloodhound trait. When the hounds of the Windsor Forest Hunt caught up with their quarry, they licked him. "When the runner is found," writes one follower of the hunt, "he is not torn to pieces, but greeted by dozens of wet tongues! A wonderful sight!" The hunt ended not in blood but in slobber.

But the breed's hellhound image endures, and back in the United States, many bloodhound owners blame Harriet Beecher Stowe. "*Uncle Tom's Cabin* gave the bloodhound such a fearsome connotation," Catherine Brey and Len Reed write in *The Complete Bloodhound,* "that to this day many people believe he is a savage and dangerous animal." Another author complains that on hearing the word "bloodhound," "nine people out of ten conjure up a tearful scene of poor Liza . . . frantically scrambling across the ice floes with a pack of vicious, red-eyed, froth-drooling beasts baying sadistically at her heels."

As it happens, these authors are wrong to malign Stowe. Bloodhounds did, of course, chase Liza in the 1927 movie of *Uncle Tom's Cabin* and in innumerable nineteenth-century stage productions. They became such a standby that a humorist exiting one unfortunate staging remarked, "The dogs was good, but they lacked support." But Stowe's version of Liza's crossing does not include any dogs at all, and her book never once mentions bloodhounds.

The real problem for the bloodhound is not, after all, its appearance or its name or Harriet Beecher Stowe, but the whole troubling question of scent, which human beings are ill-equipped to understand. To find out more, I took a trip up to East Greenbush,

New York, to visit with David and Hilda Onderdonk. The Onderdonks were soon reciting how they often travel long distances to help in a search, only to have their hounds' findings ignored or even ridiculed. In one case of an eighteen-year-old man reported missing, they drove to Bennington, Vermont, and put a hound on a trail that was then eight days old. According to the Onderdonks, the hound picked up the man's scent on bushes in a residential area and followed the trail into a grocery store and a bank, sniffing high up on storefronts and other buildings. He stopped finally at a bench outside the bus terminal. But the man's family sent the hound packing, disbelieving him not just because the trail was old and ran through impossibly busy streets, but also because investigators already knew that no one in the terminal had sold the man a ticket. This turned out to be correct. The man later phoned home from California and said that he had never gone into the terminal, but bought his ticket on the bus, for which he waited on the bench outside.

This sort of anecdote inevitably elicits the "Come, come" response from scientists. But Hilda Onderdonk put the case for believing the bloodhound this way: "You buy a coonhound to run coons. You buy a beagle to chase rabbits. You buy a bird hound to flush out birds. So why do you find it so hard to believe that bloodhounds can find people?"

Three likely reasons sprang to mind: People don't want to think that they are smelly. They particularly don't want to think that they smell different from everybody else. And if indeed they *must* smell, they don't want to think that, even after vigorous bathing, they leave an odiferous wake lingering behind them for days at a time, as if they were no better than a raccoon fresh out of the garbage can. More to the point, humans cannot smell whatever it is about them that the bloodhound smells. And because this scent, this aura, is imperceptible, it also seems implausible.

In the human nose, all of the olfactory cells, which do the work of smelling, are located high up in the nostrils in an area less than one inch long. In a bloodhound, the drawn-out canine snout accommodates an olfactory membrane that is at least six times as long. But length only begins to suggest what is really an exponential increase in sensitivity. The olfactory membrane in all mammals is spread out across tiny, paper-thin bones rolled up in scrolls, called nasal conchae. "The more scrolls, the larger the surface area for the olfactory membrane," says Rosemarie Williams, a veterinary neuropathologist at Tufts University, "and that's factor number one. The second factor is that this membrane contains individual olfactory cells, and each cell has little cilia, or hairlike fronds, that project from it. They're actually extrusions of the cell, to increase its surface area."

In humans, the scrolls and cilia translate into about three square centimeters of olfactory membrane. But in a bloodhound's heavy snout, with the dignified Roman bump accommodating more conchae and thus more smelling power, there may be as much as 150 square centimeters of membrane. Humans also have more highly developed mechanisms for shutting off the sense of smell. In humans and bloodhounds alike, the olfactory cells rapidly get used to a new smell; they adapt their electrical polarity so that they stop sending out nerve impulses unless the concentration of scent increases. This is part of the reason a smell that is overpowering on entering a room seems almost unnoticeable a minute later. In addition, humans have evolved a powerful mechanism in the brain to suppress the conscious recognition of an odor. "We need to direct our attention toward other things," says Williams. "This adaptation that we have evolved is one reason that we are a more intelligent animal."

But intelligence is a highly relative concept. An animal that needs its nose to locate prey might be fatally handicapped by these same mechanisms. A bloodhound thus does not have a strong suppression mechanism; it needs to continue consciously following the scent until it finds its quarry. It also needs to avoid olfactory fatigue—the adaptation in electrical polarity—so as not to miss minute particles of scent on the trail. Fortunately, it is possible to revive the olfactory cells with short, sharp bursts of air, alternated with momentary pauses during which the nose is emptied and the olfactory cells rest. Humans (even wine tasters) do this ineptly; they need to use their diaphragms to sniff. Bloodhounds and most other mammals use just the muscles located in their nostrils and pharynx.

Bloodhounds are biologically adapted for trailing their prey in one further regard. The process by which the nose recognizes an odor is not fully understood, but there are apparently specific receptor sites for specific odors. In one explanation, recognition occurs when a scent molecule fits into its corresponding receptor site, like a key into a lock, causing a mechanical or chemical change in the cell. Bloodhounds apparently have denser concentrations of receptor sites tuned to human scents. These traits no doubt occurred fortuitously in a few animals early on. By breeding these animals together, and then breeding their offspring back to them, people have engineered a breed in which this highly specialized talent has become dominant.

This brings us, after much hemming and hawing, to the real point: When a bloodhound trails a human being, what does it actually smell? "Pull out the front of your shirt," Hilda Onderdonk ordered, as the Rensselaer County Search and Rescue truck lurched and bumped through the streets of East Greenbush. "Do you feel hot air rising?" The Onderdonks regularly lecture schoolchildren about bloodhounds, and they have this lesson down pat. "Your skin is 98.6 degrees," she said, "and the air next to your body is also being heated to 98.6 degrees. Now, your body is constantly shedding skin, and that hot air brings up the bits of skin and the bacteria that feed on them. These little pieces of skin are called 'rafts.'"

"They come out over your head," Onderdonk continued. "Then they get cooled and they fall. It's like an invisible dandruff, and it falls in an umbrella effect around your feet. If there's any breeze, it'll get blown away. But sooner or later it gets caught in the grass or the bushes. This is a scent pattern, the way it falls, and a bloodhound trails that scent."

The human body, which consists of about sixty trillion living cells, sheds exposed skin at a rate of fifty million cells a day. So even a trail that has been dispersed by breezes may still seem rich to a bloodhound. The body also produces about thirty-one to fifty ounces of sweat a day. Neither this fluid nor the shed skin cells have much odor by themselves, but the bacteria working on both substances are another matter. In his book *Life on Man,* microbiologist Theodor Rosebury estimated the resident population of a clean square centimeter of skin on the human shoulder at "multiples of a million." As they go about their daily business breaking down lipids, or fatty substances, on the skin, these bacteria release volatile substances that strike the bloodhound nose as a whole constellation of distinctive scents.

Diet, bathing habits, toiletries, the home environment, and other factors can all influence the number and variety of bacteria on the skin and also an individual's scent. People from the same family not only share the same basic genetic constitution, but also the same soap or food or laundry detergent and will thus naturally have a similar scent. This can sometimes be confusing for a bloodhound, according to David Onderdonk—for instance, when a child is missing and the family has just crisscrossed the neighborhood in a futile search. Still, everyone's scent, or constellation of scents, strikes the bloodhound nose as distinctly individual; it is another sort of fingerprint. The bloodhound has been bred to discriminate even among the subtle differences within a family, and in the end this is what makes the breed most remarkable.

To demonstrate, the Onderdonks had arranged for a local police officer and his twin daughters to meet the rescue truck at a nearby golf course. En route, the Onderdonks' dogs were utterly blasé about the challenge ahead of them.

Indeed, they were sleeping in their cages, each of them a heap of dewlap and flews dreaming on its own double cushion of foam rubber, with a personal fan nearby for added comfort. Meanwhile, the human rescue workers sat on uncushioned plywood in the back, with no fans, bouncing across the potholes and sweatily throwing off volatile substances. Hilda Onderdonk was unapologetic about the disparity. "Could *you* find the victim?" she inquired.

In truth, this disparity extended to the Onderdonks' whole way of life. They refer to themselves as guests in their own home and to their home as a large doghouse. Their front door is boarded over and barred to human access, but the two bloodhounds can come and go at will via a hatchway leading to the fenced-in yard. In the bathroom, the roll of toilet paper, a favorite bloodhound plaything, is hidden in a closet. David Onderdonk started to hide his dentures at night after waking up to find a bloodhound named Mountain Dew staring gleefully down at him with a mouth full of human uppers. The Onderdonks have owned bloodhounds for fifteen years, having got the idea much earlier when one of their six children disappeared briefly. Now that the children are grown, the elder Onderdonks said the rescue work gives meaning to their lives. They see the dogs as extensions of themselves, or vice versa. On the trail, David Onderdonk, who weighs just 130 pounds, stumps along behind his dog as if the two of them are permanently attached: "I've been pulled over a cliff. I've had my head split open. I've got scars all over from being banged into things." A bloodhound on the trail, he said, is very single-minded. Bloodhound owners tend to be the same way.

At the top of the fairway, Onderdonk gave instructions to the two fourteen-year-old girls. These twins were so identical, according to their father, that as infants they had to go back to the hospital several times to have their names straightened out according to their footprints. One of them must now wear her hair short, the other long, so their parents can tell them apart and also to prevent them from swapping places in school. According to their mother, they do everything else, down to perfume and deodorant, alike.

The long-haired girl left behind a nightshirt as a scent article for the dog, and the two of them struck out across the golf course for a hundred yards. Then they split, with long-hair heading out of sight on the left. The hound named Moonshine now came out of his cage, looking semiconscious. Onderdonk put on his harness, offered the hound the scent article and gave him the command: "Find 'em." In an instant he was being dragged across the fairways by a hound who was completely indifferent to the scent trails of twosomes and foursomes that had been hacking their way across this landscape all day. Moonshine was unmistakably on the trail of the two twins, and at the split he headed left at a lope, without a moment's hesitation. He quickly overtook the long-haired twin, and gave her a quick confirming sniff of recognition. Then, just as suddenly as he began, Moonshine lapsed back into happy indolence.

Scent obviously has a power humans can never fully understand. For us, no sense plays more evocatively across the mind or has as quick a hold on the emotions. Yet it is a language that we perceive only distantly and incompletely, and one we perhaps suppress because we vaguely intimate its force. The bloodhound meanwhile inhales every nuance. It reads us better than we read ourselves, better than a mother reads her own child. Is it any wonder some people think bloodhounds are spooky?

Looking for
Mr. Griz

The previous night, a grizzly bear named 104 had killed an elk calf hereabouts, out near the east gate of Yellowstone National Park. The signal from her radio collar indicated she was still in the area, but it wasn't possible to pin down the location from the sporadic clicking on a tracking receiver. "Well, shucky darn," a researcher named Steve French commented. She was probably bedded down for the day with her three cubs in the trees above the road. We headed below, picking our way across a rough slope, between them and their most recent ungulate appetizer. Someone muttered uneasily about how touchy a sow with cubs can be toward intruders.

Marilynn French, Steve's wife and research partner, had spent three hours the night before watching 104 graze with studied nonchalance along this slope until the bear pounced on the elk calf. All that remained now were four lower legs and a spotted hide dressed out as cleanly as if a tannery had done the job. A researcher from the Interagency Grizzly Bear Study Team made notes in a little yellow pad. "You call that seventy-five percent?" he inquired, indicating the carcass with a pencil.

"I call that a hundred percent," said French. "They licked the plate."

The government researcher raised his eyebrows in rueful assent and seized the opportunity to scan the hillside for movement.

"Calvarium," French remarked, picking up a splinter of skull. "Elk brain must be a real delicacy." He began to expound on the grizzly's dining habits, in an odd mix of Texas country boy dialect and precise anatomical detail, with maybe just a tad too much of the latter for the average person's taste, meanwhile scouting through the trees to locate the flattened grass where the sow had bedded the night before and the turned-up dirt where it had dug licorice root with its finger-length claws. "This is a magic place for these guys," he whispered. "They had cover. They had drinking water. They was eatin' elk. It was cool."

Back at the carcass, the government researcher was collecting scat samples in plastic bags, less enthusiastically. "This is what bear work is all about," he remarked. "Being scared and picking up shit."

On the subject of being scared, the grizzly research world is in unanimous agreement. On almost any other subject even remotely connected to grizzly bears, however, there is fierce dispute. This includes, above all, the question of what bear work is all about, which is why I had come west to spend time with the Frenches.

As practiced by professional scientists, grizzly bear work nowadays is mainly about numbers: While the grizzly population in western Canada and Alaska remains healthy, fewer than a thousand grizzly bears remain in the Lower 48 states. They are confined to just 1 percent of their old territory, which formerly stretched from Ohio to the California coast. Most now live along the Continental Divide in northern Montana, with a concentration around

Glacier National Park. A smaller population of about two hundred bears survives in apparent isolation in the Yellowstone area, several hundred miles to the south. The federal government listed both populations as threatened in 1975, and much of the scientific number work since then has had to do with whether these populations are recovering and when they can be safely "de-listed." To find out, researchers trap, tranquilize, ear-tag, lip-tattoo, radio-collar, and radio-track bears. They also employ abstruse demographic tools like "congruence analysis," a "cumulative effects model," a "habitat quality index," even "volumetric analysis of grizzly bear scat by month," all to log numbers on a computer, electronically crunch them, and figure out whether the government is doing a good job managing bears.

By contrast, the Frenches, who are amateur naturalists, merely watch bears, something modern scientists rarely have time to do. They also film them. They are throwbacks to the nineteenth century, when numbers mattered less to naturalists than what they could see, and the most interesting work often came from diligent but uncredentialed observers in the field. In the case of the Frenches, fifteen years of grizzly work have helped scientists appreciate little niceties of bear behavior, like how fond grizzlies have become of elk calves, or how they like to disappear to the mountaintops in August and gorge themselves on rich pockets of estivating moths.

This tradition of straightforward field observation has revived in recent decades, most notably in the person of the late Dian Fossey, and the Frenches view themselves as following in her footsteps—with the small caveat that the grizzly, unlike the gorilla, really does eat people. Unlike Fossey, however, the Frenches generally watch their subjects from a distance, partly out of self-preservation, but mainly to avoid displacing them from their limited habitat. They also avoid politics. In the disputatious world of the grizzly, where the egos tend to be as big and quarrelsome as the bears, this isn't as simple as it sounds.

• • •

French is a short, athletic, neo–mountain man, given to chewing tobacco and spitting it into an empty can of diet soda. He was wearing a battered stetson when we met, over scraggly brown hair, bright blue eyes, and a beard like the bow fender on a tugboat. In their lectures, Marilynn, who has the lean build and the understated manner of a long-distance runner, narrates the basic facts about bear behavior; Steve handles the color commentary. She describes their observation that earthworms tend to bunch up under tufted hairgrass in wet weather. He gets to talk about how the grizzlies "flip 'em over and, *thp-thp-thp,* it's like spaghetti."

Until he takes off the stetson and puts his black reading glasses low down on the bridge of his nose, you would not guess by looking at him that he is a physician trained in surgery. Nor does he exactly want you to know that for six months of the year he is director of emergency medical services at Evanston Regional Hospital in Wyoming. "I'm a Texas truck driver's son who happens to have a doctor's degree," he said. Attitudes in and around his chosen profession make him edgy. "I wasn't raised to be an asshole."

French, now age fifty, got interested in grizzlies as a surgical resident at the University of Utah in the 1970s, while patching together people bears had taken apart. "The first time we had a patient that had been chawed on by a griz, I thought, 'What the hell is this?' If the guy had been beaten up with a pool cue I wouldn'a thought anything of it. But what kind of animal does this?"

On vacations, he and Marilynn drove up to Yellowstone, and like other tourists familiar with the proverbial association between bears and woods, they wasted a lot of time looking in barren thickets of lodgepole pine without ever seeing a grizzly. Other interests intervened. French weighed 226 pounds in 1977 when he got disgusted with being a fat physician and started to run. He quickly became good enough to clock a 2:18:40 marathon and compete in the 1980 Olympic Trials. Marilynn had gone back to school and was also running, on the University of Utah's national championship cross-country team.

Then she got pregnant, and soon after, on a camping trip, she discovered that a canyon where her family had hiked for generations had been bulldozed, dammed, and turned into a reservoir. "I saw how quickly things like that can happen," she said. Sending off checks to environmental groups with large offices in Washington, D.C., no longer seemed adequate. "I wanted some place I could have memories with my daughter," she said. The Yellowstone wilderness seemed permanent, and the study of grizzly bears sounded, in 1983, like one good cause worth a lifetime of effort. At that point, a lot of people, including the Frenches, feared that the grizzlies for which Yellowstone was famous were mostly dead.

In the summer of 1970, Yellowstone had shut down its open-pit dumps, abruptly cutting the bears off from what had been their main source of food for more than a half-century. The motive was sound: to return fat, Twinkie-dependent bears to a more natural way of life, in accord with government policy that a national park should function as an ecosystem, not a zoo. In park jargon, the bears then were "habituated" to human presence, and "conditioned" to human food. Park rangers hoped that breaking this connection would get the bears away from roads, where bear-jams and maulings were a perennial problem. The decision touched off one of the bitterest disputes the environmental community has ever known, and it remains the seminal political question for the Frenches and every other student of the grizzly.

From the start, the Park Service faced vocal opposition from Frank and John Craighead, independent researchers who had pioneered radio-tracking and other techniques in a study of the park's dump bears. They predicted that closing the dumps would cause grizzlies to starve and become extinct from Yellowstone. As

Grizzly Bear

grizzlies dispersed from the dump sites, the Park Service began to kill or remove bears that continued to seek human food, and other conservationists quickly joined the protest. By the Park Service's own count, it eliminated 83 grizzlies in 1970–1971, a huge percentage from a genetically isolated population later estimated at 234 bears. Steeped in that era's deep mistrust of government, many people believed the unreported losses were much higher. Critics denounced it as "an extermination campaign" conceived to make even this last vestige of grizzly country safe for people.

The idea that the Yellowstone grizzly had been permanently subjugated, if not destroyed, became a set piece of environmentalist lore. The Frenches, who had read Frank Craighead's book *Track of the Grizzly*, an account of the debacle, generally accepted this grim view. But as they accumulated seasons in the field, and as the bear population recovered during the 1980s, they began to think instead that the closing of the dumps may ultimately have set the grizzlies free. "The dump bears were like overfed pigs at the trough," said Steve, and it affected both their appearance and their behavior.

For example, in the dump days, when up to ninety grizzlies a night brawled around a single dump, researchers believed grizzlies to have minimal courtship behavior. It was common for bears to move unceremoniously from one partner to another in the course of a night. But in the wild, said Marilynn, "We didn't find that. Bears spent weeks together going through this very complex, ritualized courting relationship."

Here Steve cut in with color commentary: "He comes up very aggressively looking to mount her, and she physically rejects him. She'll cuff him and knock the shit out of him, and he'll take it. He just watches her and follows her around. Over the course of days, she begins to tolerate him and allow him to get closer, maybe lay his head across her rump. If he gets too aggressive, she'll turn around and whack him. But eventually they become very affectionate and cuddly."

"You see very big adults playing just like cubs," said Marilynn. "They'll be standing on their feet and pawing each other, or rolling around on the grass like when they were little."

The inference the Frenches drew from such observations was that Yellowstone's grizzlies might be better off now, not a popular idea in the West, where it is well known that the federal government never does anything right, not even accidentally. Despite their aversion to politics, this idea put the Frenches on a collision course with widespread common knowledge that the bears were still out there starving—common knowledge vividly embodied in the person of Doug Peacock, an amateur grizzly watcher of far greater renown than the Frenches. Peacock has done more than anyone to keep alive the memory of the grizzly debacle of the 1970s. Even now, he likens what the government did to Yellowstone's grizzlies to what it had done to the villages of the Blackfeet and the Vietnamese. To Peacock, suggesting that the bears might be better off now is like suggesting the Indians were better off for Wounded Knee.

This is, moreover, a collision in which popular opinion appears stacked against the Frenches, who remain unknown outside the grizzly research world. Peacock, on the other hand, is a fixture of western folklore. He was the model for Hayduke, the beery, foul-mouthed saboteur, bane of all developers, at the heart of Edward Abbey's cult novel *The Monkey Wrench Gang* and its sequel, *Hayduke Lives*. He has also chronicled his own experiences among grizzlies in a short film, *Peacock's War*, and in an autobiography, *Grizzly Years*. The writer Peter Matthiessen has described him as having "more experience with wild grizzlies . . . than anyone alive." As I thought about the Yellowstone grizzlies, it seemed to me that Peacock's life among the bears provided a shadowy counterpoint to the Frenches'.

Peacock began to study grizzlies at Yellowstone and Glacier in the mid-1970s at about the same time Steve French was sewing up bear-mauled campers in Salt Lake City and on short stints in the Yellowstone Park hospital. But where French readily admits that he came to Yellowstone expecting something like Six Flags over Texas, Peacock already knew wilderness and he knew how to find bears.

He grew up "running loose in the North Woods" of Michigan, learning not just wilderness, but Indian lore. As an amateur archaeologist at the age of fourteen, he made a major find of four-thousand-year-old Indian corpses wrapped in beaver pelts. He also got drunk for the first time and killed fourteen pigs in a shooting spree. He studied geology at the University of Michigan, but dropped out and headed west. In his book, he depicts himself in the late 1960s returning to the American wilderness and its fiercest incarnation, the grizzly, as a kind of catharsis, a way of forgetting what he had seen during a two-year tour patching up soldiers and children as a Green Beret medic in Vietnam. The grizzly was also a way of reviving the war's intensity of feeling, "the naked authenticity of living or dying."

With the idea of talking about that era in Yellowstone and about the Frenches, I looked up Peacock at his home on the outskirts of Tucson, where a Park Service sign in the front yard asserts, "Remain At Your Car All The Time," and a bumper sticker in the window says, "Developers Go Build in Hell." I was expecting Hayduke, whom the late Ed Abbey, Peacock's closest friend, characterized in *The Monkey Wrench Gang* as "a manic-depressive psychopath," sullen, inarticulate, and sodden with beer. Apart from extraordinary native sense in the wild, and an interesting penchant for reducing heavy construction machinery to scrap, Hayduke seemed like a one-dimensional sort of folk hero.

"Hayduke was a dolt," said Peacock, who plainly wasn't. He turned out, in truth, to be a complex, intelligent, and highly likable character. Physically, he looked like a bear himself, built low and thick, suggesting considerable power, and with a broad, dish-like face, a grizzled beard, and milky brown eyes. He grinned readily, talked loud, and was as critical of himself as of anyone else—with the exception of the Park Service and, by extension, the Frenches, whom he plainly despised, describing them in his kinder moments as apologists for Yellowstone. For him, wilderness was the cardinal value, and Yellowstone was a domesticated environment, "a little more than a zoo but a little less than an ecosystem, or anything wild."

Wildness also mattered personally. He despised bureaucracies, describing himself as an outlaw. His preferred style was to slip through the shadows of society and disappear into the woods like a guerrilla. The Frenches, by contrast, work closely with government researchers and describe themselves unromantically as "GS-Zeros."

In his book, after fulminating about the events of the 1970s, which continue to outrage him, Peacock draws one original conclusion about the bear itself. Yellowstone is not exterminating its grizzlies, he writes. But by continuing to remove troublemaking bears, even at the present greatly reduced level, it is domesticating them: "The predatory segment of the population had probably been killed off selectively, and continues to be culled . . . because predatory bears are bolder and more visible." The end result, he suggests, will be "an animal who looks like a grizzly, albeit a small one, but whose behavior will more closely resemble that of the meeker black bear." For Peacock, the Frenches are collaborators in this process.

His conclusion turns out, however, to be wrong.

In the summer of 1983, Steve and Marilynn French put in their first long stint of serious bear watching at Yellowstone, and what they began to document, at a time when hardly anybody else was looking, was the emergence of a grizzly fundamentally different from its predecessors at the dump. Number-crunching scientists knew the grizzly population was much younger, which was unsurprising after the "control actions" of the 1970s. The bears also weighed less, by a third to a half, and had a lower reproductive rate, which seemed to support the contention of critics that they were starving. In 1983, Frank Craighead was still urging the park to kill elk and hand out the carcasses to save the bears.

The Frenches steered clear of this spat and set out on what they conceived of as an educational mission, to "deliver the heart of the grizzly bear in grizzly country" to the general public, unobscured by politics. If Peacock's approach to the grizzly was instinctive and characteristically haphazard, the Frenches' was utterly methodical: They took a weeklong course at the Yellowstone Institute to find out what they'd been doing wrong over the previous five years. Then they threw themselves into the grizzly world with the same intensity that gets a fat physician from heart-attack weight to a 2:18 marathon in under three years. Their work soon earned the

attention of Yellowstone's research community, which generally
resists independent researchers. The Frenches were witnessing the
reeducation of the grizzly bear as a predator—and this was news
the Park Service was eager to hear.

The closing of the dumps in 1970 still seemed to critics like
disastrously bad timing for the bears: The elk population then
was artificially low as a result of old-style park management, and
the native cutthroat trout were just beginning a comeback. But as
these other resources recovered, the Frenches documented how
quickly the "new" grizzly was adapting. Their efforts to witness
bear behavior and record it on film and in detailed field reports
sometimes went beyond what salaried researchers might have
done. During the fires of 1988, the Frenches stayed on so long listening to two radio-collared grizzlies on their receiver that Steve
thought his truck was about to explode in the advancing flames.
(It didn't, and the two grizzlies were back the next day working
the burned-over meadow.) Another time, he spent three days in a
tree, standing on a narrow, nailed-up crosspiece and sleeping in a
"wall womb" hammock, to see how grizzlies fished in spawning
season. Together, he and Marilynn watched a single bear for ten
days and recorded that it killed an average of ninety-eight cutthroat trout a day. One bear killed twenty-four fish in twenty-
eight minutes.

From late April into September, the Frenches now spend about
ten hours a day, seven days a week, in the field, starting out at five
in the morning and again around five in the afternoon. They average several grizzlies a day. In between they write up their field
reports, run, play softball with their daughter McKenzie, and
sleep. If the Frenches are not number-crunchers, their field work
has nonetheless become highly scientific. Among other things,
they now record air temperature, barometric pressure, time of
day, behavior, and location by Universal Tranverse Mercator ticks
on a topographic map.

On the back of a typical field report, a hand-drawn map of
one observation looks like a football play, all x's and arrows.

Each x represents a place where the bear dug up a pocket gopher cache, which can contain up to two quarts of roots and bulbs and also, with luck, a pocket gopher. "I've seen 'em dig a trench twenty feet long just to get this one little gooey gopher," said Steve. "I can't perceive that it's worth the energy, but they get all excited when they hear that little guy squeak. It's kind of like a Twinkie." The Yellowstone grizzly, he said, is an intelligent, inquisitive, highly adaptive animal, and no longer needs real Twinkies. It is doing fine on a remarkably diverse assortment of natural foods, and for grizzlies as for fat physicians, thinner may sometimes be better.

One evening we hiked out into a broad grassy valley popular with grizzlies. Steve and Marilynn were backpacking about fifty pounds of spotting scopes and camera equipment each, which we set up on a knoll overlooking scores of grazing elk. "We pay attention to what the elk are doing," said Steve. "There are more of them than there are of us, and they have a vested interest in looking for grizzlies." This is especially so in June when the elk are just finished calving.

The elk cow's harrowing survival strategy is to bed down her nursling for the day in the sagebrush meadows, and walk away. The calves would be too easy to spot if they stayed with their mothers, and for their first month, they are too wobbly to flee a predator. The calf must lie still and rely on the camouflage of its speckled coat in the hope that a grizzly bear crisscrossing the area like a bird dog will not see it and snatch it up.

A small point of clarification: We were of course rooting for the grizzlies. Elk calves are common as pennies in Yellowstone, and they now serve, said Steve, in roughly the capacity of "a Bobo platter in a Chinese restaurant. Easy pickins. Little bites." He looked out over the blank landscape and remarked, with a hint of vicarious hunger, "I would predict that there are thirty elk calves bedded down in that sagebrush meadow."

Alas, the grizzlies did not take the bait, and for our bloody-minded thoughts we were duly punished. In the middle of a meadow on the hike out, a huge bison rolled straight at us, snorting malevolently. There were no trees to climb or rocks behind which to cower, so we discreetly accelerated, and the bison eventually veered off. "I swear, I know we are more concerned with bison than with bears, any day," said Steve, and from a relatively safe distance, he yelled back, "*Yo' mama be a ruminant!* He's probably lowest on the bison totem pole, so he had to push us around and show he was dominant. Now he can walk back and say, 'I be *bad.*' "

Over the years, the Frenches have frequently been run up trees by moose or bison. They have also been caught on a mountaintop in a thunderstorm when the lightning was clapping down around them and they had to squinch among the rocks and plug their ears to avoid being deafened. They have never, however, been treed by a grizzly. Climbing a tree to escape a charging grizzly ranks with such other not-so-bright ideas as the anti-bear spray now available in what French calls "the piss 'em off" size; a charging grizzly is just too fast. Better to fall on one's face, cover the back of the neck, and lie still. In any case, close calls don't figure largely in the Frenches' reputation, and this turns out to be one of the more intriguing political questions raised by their work.

Though they also spend time in the backcountry, the Frenches do some of their best bear watching from one of the most heavily traveled roads in Yellowstone, overlooking Antelope Creek. They pulled in at five one morning during my visit, coasting downhill to their viewing point with the engine off. For the next two hours, they talked in whispers lest they disturb bears out on the ridges below them. When a tourist stopped and let a yapping dog out of the car, Steve quietly suggested that someone strangle it. Grizzly bears are sensitive to every new human intrusion, including lap dogs and scientific research. What they do not see, he added, they often smell. He was once keeping his distance and watching a bear muzzle-deep in carrion when the wind shifted and the bear lifted its nose and turned to stare straight at him. Grizzlies generally take such hints to flee before people get close: "The first thing you know is the flash of butt running away from you. And you're going *huh-huh-huh*," French panted, miming terror. "It's already over before it gets your attention. But it still gets your attention." As the day lengthened, tourists in Winnebagoes whipped past and sometimes stopped to ask questions. "Elks and grizzlies?" one man mused. "Are they compatible?" Unperturbed, French peered intently through his spotting scope at a goosed-up elk on a distant ridge, followed its gaze to the treeline, and at 7:25 said, "I got a griz." The distance turned out to be an un-macho three miles.

This is not the stuff that made Dian Fossey a movie heroine. When I visited Doug Peacock, he derided the Antelope Creek watch as "a travesty" of the tamed Yellowstone: "Watching the bears copulate through telescopes and giving them names." (In fact, the Frenches refer to bears by number, not by name.) His own methods smack more of *True Adventure:* "When he's in bear country," a friend later told me, "he carries himself as a bear would, the way he moves through the bushes, like a bear foraging. We'll be walking along and he'll be taking the point, and all of a sudden he'll stop, either point to his nose or point to his ears, and after a couple of minutes, there's a bear. He can sense their presence." But there was more than that, said the friend: "I don't know what it is with Peacock. It's like he's got this innate trait with bears where they seem to be tolerant of him." This ursine aura, and the sense of personal connection with individual grizzlies, is at the heart of Peacock's underground celebrity in the West. It is also the central flaw in this method, as it was for Dian Fossey, too.

Like the Frenches, Peacock argues that what the grizzly needs most is to be left alone. He has nothing good to say about ear-tagging, lip-tattooing, and other techniques of demographically oriented biologists, which to his mind serve mainly to habituate the bears. He wages an informal and sporadic campaign to protect grizzlies from human encroachments. At Glacier National Park, he forced federal authorities to prosecute a helicopter pilot who buzzed the bears for the pleasure of tourists, and he admits to poaching livestock on public land, where they are a temptation to grizzlies and an excuse for grizzly control.

But what Peacock cannot seem to do is leave the bears alone himself. He is drawn among them, like Fossey sitting among her gorillas, and he reports that they generally treat him "much as they did other, more dominant bears." Sometimes, with his head cocked to one side, he talks a bear out of charging. Sometimes he stares a bear down. His stories have to do less with how bears react to their changing environment than how they react to Doug Peacock.

What the Frenches give us, by keeping a discreet distance from the bear, is a portrait of an American animal, in the wild.

These differences may not at first seem to matter much in the context of the Yellowstone grizzly's future, which continues to lie not with the Frenches or Peacock, but with the number-crunchers and their cumulative-effects model. The model allows demographers to assign a quantitative value to every proposed new development in grizzly country, in terms of the effect it will have on the population of breeding females. This number is typically a fraction of a bear per year, allowing each developer to tout his project as relatively harmless. But fractions have a way of adding up. I sat down with John Varley, the park's chief of research, to review the developments now being built or proposed in the greater Yellowstone ecosystem, including expanded tourist accommodations within the park itself. Varley saw little room for long-term optimism. The cumulative effects are adding up, he said, to the point where society must now choose whether it wants grizzly bears at Yellowstone or new hotels. Without some change in direction, he predicted that the bear population would begin to decline again within the decade.

The Frenches were equally pessimistic. One of the new developments under consideration at the time of my visit, for example, was a gold mine expected to bring six hundred people to Cooke City just north of the park. "Everybody's going to kick and scream," Steve French said. But society loves a compromise. "In the end they'll come up with some environmental mitigation and call it a win-win situation." He paused. "You don't have a win-win situation when you put six hundred people into grizzly habitat. Three years down the road, it's going to be something else over here. *Win-win.* And then it'll be *win-win* over here, and *win-win* here. And each one taken separately will be insignificant to the grizzly bear and its long-term survival. But we will win-win the bear to death."

In 1997 the U.S. government ultimately found just such a win-win solution to the Cooke City mine proposal, persuading a Canadian company to give up its claim to the land in exchange for $65 million worth of federal land elsewhere. "Is it going to cost us $65 million every time we resolve a grizzly habitat conflict?" French asked afterward. "If so, we can't afford it."

Here, however, one of Peacock's criticisms of the Frenches sticks. Peacock regards it as an environmentalist's obligation to resist such development, to oppose the removal of habituated bears, to blow the whistle on concessionaires who regard Yellowstone as a profit center and operate in flagrant disregard for park wildlife. (Bears continue to become habituated in part because of negligent handling of food and waste by park restaurants and hotels.) These are subjects on which the Frenches also have strong feelings, and yet they keep silent. They are disciplined enough about abstaining from politics that they will not even talk about Peacock on the record. The difference perhaps is that where Peacock is always drifting through, the Frenches intend to spend their lives at Yellowstone. They have formed a Yellowstone Grizzly

Foundation, and they talk in terms of a thirty-year study, which depends on the support and cooperation of the Park Service. They are conscious that a previous independent research team, the Craigheads, had to go elsewhere after breaking publicly with the Park Service. Thus even the Cooke City gold mine had to be somebody else's fight. "Our drive is for conservation through long-term public education," said Steve French. "You read the government literature, and they're talking about all the things they're doing for bears. You read the conservation literature, and they're telling people what they're doing for bears. *All we're telling people about is bears.*"

Under the circumstances, it is natural to wonder whether education alone is enough. If the Frenches don't speak out, it falls to people like Peacock to kick and scream, to act outrageously on the grizzly bear's behalf. But for all his folkloric appeal, Peacock lacks the credibility to persuade society at large that at least in this one small place, our dominion over nature ought to be on hold; moreover, the Yellowstone grizzly of his imagination is already a degraded, half-domesticated specimen, hardly worth the trouble of saving. If we want to know what Peacock is kicking about, if we want to savor each small discovery about how much more marvelous the untamed grizzly bear is than another air-conditioned hotel, then it is the Frenches that society needs to hear. For them, as for Fossey and just about every other conservation scientist in the world today, the politics of survival are inescapable.

Two mornings after the bison episode, we hiked out into a sagebrush bottom in time to see a long line of elk and their calves (the bait) making an exodus from the area, a bad omen. The Frenches set up their cameras on a high point above a grassy landscape corrugated with deep draws. There was a stream ahead and, across it, a low meadow leading up to green, aspen-filled chutes among the bluffs. A group of bighorn sheep crossed a high slope opposite, and a couple of sandhill cranes made their creaky, chittering sound like wood rubbing against wood. Across the river and far off to the right, a small herd of elk grazed.

"We got to watch our flanks," said Steve, and he started to tell a story about a time when a bear inadvertently ran a herd of elk right at them from the side. Before he could finish, frightened elk bolted into view on our left flank. "Oh, heavens," he said. "What's going on? Get ready, Marilynn, there could be a bear over there."

"There's a bear, there's a bear!" she shouted back. "Right on 'em." The grizzly came into view at a rocky gallop, three-quarter speed, with its loose flanks rippling in the sunlight. Then it disappeared after the elk into a draw, direction unknown. The bear was five hundred yards away, less than thirty seconds for a grizzly.

While we waited for it to reappear, Steve digressed about a grizzly they once timed on a long run uphill in broken terrain. It had a stride of seventeen feet and a speed of thirty-two miles an hour, "and he wasn't even pissed at anybody," he said. He began to reflect in the anatomical mode: "They're plantigrade, like us; they walk on their heels. Elk are digiform; they walk on their toes. So why are bears so darn fast? Do they have incredibly fast twitch fibers? Sprinters have a high concentration of fast twitch fibers. Or does it have to do with the way the tendons cross the joints and the bones articulate?" He spat tobacco; it was a subject for future research. "If that bear comes up over the ridge," he concluded, "why don't we all bend over and kiss our asses good-bye. Because we're gonna be a helluva lot easier to catch than that fuckin' elk."

The bear reappeared on the other side of the river, shaking off water like a dog. A half-dozen elk stood nervously on a knoll. They walked straight toward the grizzly, their usual strategy being to keep it in sight at least until the chase begins. The Frenches refer to the way a whole herd will suddenly point in the same direction, even when the landscape prevents individual elk from actually seeing the bear, as networking. The grizzly ignored them and lay down in the low meadow to work on a recent kill.

"Subadult, two hundred fifty pounds, dark brown," Marilynn remarked to Steve, who was making field notes. The Frenches had watched a sow nail two calves in thirty seconds on this same spot ten days earlier, and they speculated about whether this bear was simply working over an old carcass. The bear held on with its claws and drew something up with his teeth. "I don't know," said Steve, "that's awful . . ."

". . . chewy looking," said Marilynn. "I wonder if he could've gotten it last night?" After a few minutes, a cow approached to within ten yards. "That's real typical," said Marilynn. "That's the mother. She's got to have the urge to nurse, and there's the separation. She's had that calf with her for a few weeks." Another cow moved in behind, popped up her calf from its daybed, and escorted it out on tiptoe. The grizzly continued to eat for another twenty minutes.

"He's walking," Marilynn said finally. "7:08." It developed that the bear already had two kills, but he continued to sniff through the sagebrush in search of other elk calves. "The cows are coming down to see him."

"There's a calf right above him," said Steve. "Look at 'em run; look at 'em run! He's way off to the left. The next place you're going to see him is at that cut in the river."

Calves tumbled into the river in a panic and swam across, struggling up the cut bank. On our side of the river, elk suddenly started to pour up out of the woods. The air was filled with the screaming of the calves, like seagulls, and the strangled barking of the cows. "There he is again," said Marilynn. "Coming around this way."

If the grizzlies are learning to be predators, said Steve, the elk are learning to be prey. "For a while they were like domestic cows, they thought they could just get fat and happy." But now they're developing defensive strategies. In the Frenches' film footage, three cows pull up beside a charging bear and attempt to shoulder it off the track of a fleeing calf; another trio cuts directly across a bear's path to break its visual lock on its prey. Sometimes the calf escapes. Sometimes it stumbles, or the bear butts it down with its muzzle. Even then, a cow may harass a bear enough to distract it and let the calf escape.

Steve headed off to a vantage point a hundred yards away and then came sprinting back: "He just went over there and corralled 'em. I heard some elk screaming. Can you believe this little guy or what?" The elk came racing into view, the calves struggling to remain among them. Their brown flanks glistened with sweat and they were panting, their pink mouths open, eyes wild. The herd circled up onto a knoll. In hot pursuit, the grizzly's mouth was also open, with a look of pure joy. He flew up the slope and hurtled in among them, spinning one way and then the other, snapping wildly, scattering elk like a kid playing blindman's bluff. The cows were all around him, confusing him, and the calves danced away just out of reach. Then the whole wild assembly disappeared down the far slope.

"Goddam!" Steve French yelled, from behind his movie camera. "Why couldn't I have a full roll now?"

Later, on the hike out through head-high sagebrush, a perfect spot for stumbling into a grizzly, he remarked, "You think about this. This isn't the Serengeti. This is Yellowstone, with Old Faithful, the boardwalks, the Winnebagoes, the two and a half million people." And later still he added, "One of those dump bears woulda had a heart attack."

What's Nice?
Mice.

If you had gone to the Orange County Fair in California a while ago, and picked your way past the zeppole booths and the sideshows ("Giant Steer—10,000 Hamburgers on the Hoof!"), you might by sheer good luck have met a comely music and mathematics teacher named Roxanne Fitzgerald. She was surrounded by small cages at a booth in the livestock shed, and she wore her heart on the brim of her sailor's cap, which was embroidered with the words: "What's Nice? Rats and Mice!"

Yes, and you would have met plenty of people who agreed with her, or disagreed only about which was nicer. At this particular outing of the American Fancy Rat and Mouse Association, mousers were in the majority. They had assembled, amid the champion long-eared rabbits and bantam reds, to show their prizes: hairless mice, frizzy mice, long-haired mice, Siamese satins, orange satin selfs, agouti standards—and to discuss the fine points of the mousing life.

You might, for instance, have heard Don Fredriksen, a boat carpenter, lamenting his wife's distaste for agouti mice, which have a rather mousey gray-brown color. She lets their nine-year-old son, Josiah, keep a Chinese water dragon in his bedroom, Fredriksen was saying, but agoutis "look like house mice to her and she doesn't want 'em around." You would have seen much sympathetic nodding.

And you would have heard the mice themselves discussed with a discernment, even a delectation, that was utterly foreign to the everyday "eeek-it's-a-mouse" world. Indeed, at that moment, Debbie Prosser, the show judge, held a lilac long-haired satin in her open palm as if it were a jewel. She ruffled its coat gently with a finger to check for thickness and evenness of color.

A stern, meticulous judge, she had already sent a hairless mouse away from the judging table for having unsightly tufts on the tops of its feet; she had marked down a silver tan because its nose was too pointy and its whiskers had been chewed off. But Prosser had also been sensitive. At one point, she cupped a listless mouse in her hands and held it up to her ear, as if to hear its confession. "When they have respiratory problems, you can hear them wheeze," she explained. And looking into the eyes of another contender, she said, "You're a *very* nice mouse, you just don't compete. I'm sorry."

For the mice who *do* compete, the day culminates in the best-of-show judging, when Prosser holds up her choice and declares, "That's the ideal mouse, right there." The winner belongs to young Josiah Fredriksen, and it is an agouti.

"That's great," Josiah's dad says to the small group of onlookers (his wife not among them). "He's been up on the table a couple of times before, but that's the first grand championship he's ever had." He hesitates, then inquires, "Should we clap?"

A very good question. Should we clap for the mouse? Should we clap, that is, for the *house* mouse, which is what we are talking about here, whether in plain or fancy versions, whether gutter-born or grand champion? You do not need to be a musophile . . . a sminthophile . . . a lover of mice . . . to consider the question seriously. Whether we like them or not, mice are a hit, an improbable success story, if only in that they have distributed themselves more widely across the planet than any other mammal apart from *Homo sapiens*. They have done it with short legs and bad eyes, too, and with less help from us than we generally choose to think.

The house mouse hasn't merely traveled far, it has learned to live almost everywhere, adapting its modest physical endowment to an extraordinary range of habitats. House mice live at the bottom of English coal mines and at the summit of Mauna Kea in Hawaii. They live in the coastal deserts of Peru and fifteen thousand feet up in the Andes. They dwell in frozen meat lockers, snuggling in the carcasses that feed them. In Australia, where a few centuries ago they were unknown, house mice have made themselves the most common small mammal everywhere. And they have managed to colonize an island near Antarctica where even humans are too sensible to live. They reside in the walls of this writer's house, and nest in my shredded tax records. The Bible says the meek shall inherit the earth. The house mouse is living proof that they already have.

House Mouse en Famille

So do we clap? The question is complicated by a curious paradox. While humans have lived with mice for millennia and while we have studied them in endless detail as scientific models for ourselves, we remain ignorant about mice as mice. Even the aficionados of the American Fancy Rat and Mouse Association are more interested in mouse aesthetics than in mouse behavior. Nor do more than a handful of researchers care about where they came from or how they live. Mice, one researcher told me contemptuously, "aren't real animals."

This ignorance is even odder because mice are a subject on which humans harbor deep and contradictory feelings. A mouse, like no other animal, can evoke in an instant feelings of keen empathy and cold-blooded mayhem. We think of mice as cute little critters—those anxious, hyperthyroid eyes, the fanlike ears, the nostrils sampling the air so frantically they seem to oscillate. But we also call them vermin and make a beaten path, as Ralph Waldo Emerson said, to the door of anyone who comes up with a better way to kill them. So if we are deciding whether or not to applaud the mouse's spectacular success, we had better think about the tangled ways humans view mice, and also about mice from something like a mouse's perspective.

One of the first tangles has to do with just what qualifies as a mouse. Any small, furry rodent will do, for some people, and hundreds of species fit that description. There are pocket mice, pygmy mice and fat mice, leaf-eared mice and striped-grass mice, American harvest mice and dormice. Australia naturally has kangaroo mice, but so does New Guinea, and neither species has much connection with the American jumping mouse. The Middle East has porcupine mice. Southeast Asia has marmoset mice (also known, less winningly, as Asiatic climbing rats). And in South America, chinchilla mice are trapped for their warm fur, which is used to make garment trimmings and even whole robes.

Grouping these animals together under the common rubric "mice" is a bit like mixing beavers and guinea pigs. Scientists classify them in different genera, different families, or even different suborders of Rodentia. Nor does it entirely simplify the question to talk only about house mice. In this country, for instance, deer mice are among the most abundant mammals in almost all habitats. About fifty species are native to this hemisphere and predate the arrival of humans. But they will sometimes live indoors, too, and it will afford little comfort to householders who hear them scrabbling under the floor to know that they are not, scientifically speaking, house mice. No, it is necessary to simplify the discussion further to those small furry rodents that have climbed into humanity's back pocket and made themselves our companions full-time and in all places, the members of the genus *Mus* whose name is thought to come from the Sanskrit verb "to steal."

Mus (rhymes with cuss) is the house mouse proper, and as such it has insinuated itself not just into every corner of the planet, but into our culture. The Greeks spoofed *The Iliad* in a satire about a one-day war between mice and frogs, with great orations before battle and with the gods butting in on both sides. They also built a temple to Apollo Smintheus (Apollo "the Mouser"), their protector from infestations of mice, and Pliny reported the use of white mice to augur the future. If they had augured right, they would have seen a future of mice everywhere, superhero mice, mice to whom millions of children would sing their allegiance. My generation was raised on Mickey Mouse and Mighty Mouse ("Here he comes to save the day . . ."), and my eldest son doted on Danger Mouse, the James Bond of rodents (*"Powerhouse,"* a breathy female vocalist intoned). My father grew up with Ignatz Mouse, a dyspeptic comic-strip underdog (or *übermaus*) who was always hurling bricks at his counterpart, Krazy Kat.

What all these representations say about mice is that we tend to mistake them for humans. We seem to like mice, and to empathize with them, at least in part because they are so thoroughly subject to capricious fate. One moment their lives are the picture of domestic tranquility, the very next a king snake engorges Number 34 son, or an owl pounces on Mama. The scene is particularly affecting when capricious fate arrives in the form of ourselves, as on the infamous day in November 1785 when Robert Burns drove his plow through the nest of a "Wee, sleekit, cow'rin', tim'rous beastie." Mice lead lives of chronic instability and routine disaster; they tend to remind us, when they turn up cat-mauled on the front stoop, that we do, too. Burns wrote, "The best-laid schemes o' mice and men gang aft a-gley." Or as one mouse researcher has put it: "There is something terribly familiar about the awful situation of a mouse in the world."

But beyond that, mice have pluck. They have a way of carrying on, of getting by, even as the world crumbles daily about their ears. Forced to abandon one cozy home, they quickly set up another and another—and they survive. An 1859 report mentions a barrel of biscuits that had been sealed in Aberdeen and opened in the Canadian Arctic more than fourteen months later; it contained a live mouse. Who (except the people who wanted to eat the biscuits) could help but admire the mouse's heroic persistence?

Until recently, most scientists who bothered to think about it believed that the house mouse originated in the Russian steppes, and then freeloaded its way westward, like the mouse in the biscuits, as a stowaway in the caravans of migrating humans. But it appears now not to have been that simple. Scientists say that what we call house mice are, in fact, several separate species, *Mus domesticus* and *Mus musculus* being the principal. These two species went their separate ways one million to three million years ago, and they developed their essential characteristics well before humans settled down to an agricultural way of life. They developed, in other words, as "real animals," and we need to see them that way—rather than as mere appendages to human civilization—to understand how they have spread out and colonized the world.

Mice apparently never came to a great turn in the evolutionary road where they had to throw in their lot once and for all with human beings. Commensalism seems to have occurred repeatedly, and not always permanently. Scientists use the term "fortuitous preadaption": Mice were already accustomed to some feature of the environment, such as rock crevices, which later made it easier for them to live in barns and houses. It certainly required none of the sacrifices cats made, for instance, in submitting to domestication. The ease with which house mice can go feral even now suggests that they remain much the same animals as their distant forebears who were unacquainted with *Homo sapiens*.

As to how early house mice dispersed themselves through the Old World, they may have walked. Richard Sage, who was a researcher with the University of Missouri–Columbia in the 1980s, studied the two predominant species at a hybrid zone where their ranges meet along the old border between East and West Germany. *Mus musculus* dwell from here eastward into Asia and Japan. *Mus domesticus* inhabit the Mediterranean countries and western Europe, as well as the New World, where they arrived with the colonists. Their common point of origin is unknown (though the latest mitochondrial DNA evidence suggests northern India as a likely possibility). But Sage theorizes that after the last glaciation, the two species may have repopulated the northern countries independently, *M. domesticus* coming up via Morocco and Spain and perhaps colonizing some areas on the heels of us humans. (If each generation sent pioneers a thousand meters into territory where no mouse had ever gone, and if there were a new generation every month or so, even a shortsighted, bandy-legged creature like the mouse could find its way from Spain to Denmark, say, given several thousand years.) It is even possible that in some places humans moved into the mouse's cave, rather than vice versa. Mice simply had the sense to recognize a good thing when they saw it. If they had gotten far on their own, they were to get even farther in the new worlds opened up to them by agriculture and human transportation.

But fortuitous preadaption to human company doesn't completely explain how these creatures of distant and unsavory origin got to be where they are today (that is, in the walls of my house, and probably yours, too). To understand why house mice are so successful almost everywhere, we need to know more about how they live.

Their system of social organization is, curiously, both deeply isolationist and at the same time conducive to colonizing brave new worlds. Much of our knowledge of this system comes from Peter Crowcroft, an English researcher who spent a sizable part of his career in a darkened, windowless room figuring out what the mice around him were up to. Though his ultimate business was extermination, Crowcroft liked mice. He was inclined to settle down among them on the floor, and he sometimes sat still for an extra twenty minutes while a mouse "had a nap between my trouser leg and me."

Crowcroft's working method was to introduce various combinations of mice into the room, which was set up with a grid of nest boxes on the floor, then watch what happened. In one early instance, having allowed a mouse named Bill to become established, Crowcroft introduced another male named Charlie. "I was quite unprepared," Crowcroft later wrote, "for the stark savagery with which Bill hurled himself upon Charlie in the first instant of their meeting." Far from being the "cooperative little fellows" Crowcroft had expected, mice were acting like "savage individualists."

But for all their ferocity, mouse fights seldom resulted in death; they tended to be played out according to set patterns of behavior, culminating in the dominance of one mouse over another. And just to make sure the subordinate mouse got the point, the dominant one chased him every time they met. As Crowcroft introduced additional mice, a social hierarchy developed. Charlie fit somewhere in the middle, subordinate to Bill but dominant over other mice.

Unhappily for Charlie, other researchers have since found that the important privileges of rank accrue only to the top mouse. A dominant mouse engages in urine marking, for example, far more than do subordinates, and his urine is qualitatively different. Since mice communicate largely by scent and can distinguish individuals on that basis alone at a distance up to seven inches, this sort of thing means a lot. Male mice are generally unimpressed by the urine markings of a subordinate, but they steer clear of areas marked by a dominant mouse. Females, on the other hand, hang around areas marked by a dominant male.

Researchers now regard small, hierarchical groups with a despot like Bill at the top as the typical social organization for mice. Indeed, within Crowcroft's room, several such groups occupied and defended small territories, separated by a no-man's-land. These little mouse autocracies do not have vast territorial ambitions; mice are, if anything, contractionists rather than expansionists. If food is readily available, they may confine their movements, after an exploratory stage, to an area of a few square meters. Haystacks, even side by side, typically harbor separate and isolated groups. In western Canada, distinct populations of mice inhabit small granaries as little as one meter apart, and in England, a researcher found fifteen separate house mouse communities on one farm.

However small it may be, the territory puts its mark on the mice. They travel their home range continually and come to know their routes not just by sight or smell, but internally (or kinesthetically) as a sequence of muscle movements; they will, at first, jump over an obstacle even after it has been removed. The whole group defends its borders with a ferocity seldom seen in their internal squabbles. When a dominant male is absent, pregnant females will savagely defend the group's territory. Yet juveniles and females can freely cross into another group's territory (they lack an aggression-eliciting factor found in the urine of adult males), while an adult male trying to join a neighboring group risks death. (He is unwelcome for good reason. The usurper mates with females in the group, and ejaculation stimulates him to kill their infants. This infanticidal impulse persists for about three weeks after mating, until about the time the male's own offspring are due to be born. Then he becomes as nurturing as a lactating female.)

At first glance, none of this would suggest that the mouse—this bickering homebody—is well suited to colonizing new worlds. What makes it even more improbable is that mice combine isolation with inbreeding, and somehow parlay this apparent recipe for feeblemindedness and debilitation into an evolutionary advantage.

Because the dominant mouse in a group does most of the mating, and then mates even more frequently with his female offspring, his genetic traits spread through the population. This process quickly results in neighboring communities of mice that are as distinct genetically as two human tribes. With multiple tribes on a single farm (or, though one does not like to think it, in a single house), some are bound to display advantageous traits. They may be able to survive catastrophic cold, or elude a better mousetrap, or grow fat and sleek on a new and short-lived food source. Inbreeding thus facilitates rapid evolution. Nest-building behavior, for example, appears to have a genetic basis, and in the relatively short time house mice have lived in North America, genetic variability has enabled them to adapt that behavior to suit different living conditions: When raised in an identical environment, the offspring of Maine mice consistently use more cotton in their nests than do the offspring of Floridians. Paul K. Anderson of the University of Calgary has referred to this adaptiveness as "evolutionary serendipity." No matter what the tune, no matter if the orchestra strikes up a waltz or a jig, a minuet or a tango, there is probably a mouse somewhere who can dance to it.

Mice also of course reproduce at allegro speed. Females normally give birth for the first time at six to eight weeks of age, and every four weeks thereafter. By one researcher's estimate, a pair of mice producing six pups per litter, and with successive generations doing likewise, would yield 2,688 living animals at the end of six months. (Fortunately, this theoretical figure does not reflect the high mortality rate outside the laboratory.) It appears that their fecundity confuses even the mice themselves. Females may nurse the young of other females; even immature animals that have never given birth sometimes get stimulated into lactation and join the group effort. One can imagine the original couple overwhelmed with progeny, like octogenarian humans proud but also a little horrified to see all their grandchildren and great-grandchildren assembled in one place. Sooner or later, somebody—everybody—had better move out.

And this is in fact what happens. In a study of feral house mice living on Great Gull Island off the Connecticut coast, Anderson estimated that three-quarters of all offspring become colonizers. Sheer numbers—combined with limited opportunity in the hierarchical social world of their home territory—drive young and subordinate mice out into the world. Because neighboring groups are closed to males, they seek new territories of their own—to which genetic variability may leave them uniquely predisposed—and try to establish new communities. And thus do mice make the world their dominion.

Blitzkrieg reproduction, while crucial to the ubiquitous success of the house mice, is also the point on which it ceases to be a cute little critter and slides inexorably into the realm of vermin. Mouse reproduction is a daunting phenomenon even for the members of the American Fancy Rat and Mouse Association. You get a sense of this from the perfunctory names they give their prize specimens: AJA, ASTA, ABBA. The imagination plainly balks at the task of naming even the pick of the litter.

And as for the runts . . . Among those at the Orange County Fairgrounds who uphold the slogan "What's Nice? Rats and Mice!" there seem to be a disproportionate number of herpetologists. There are even mouse entrepreneurs, who sell what are termed "feeder" mice to zoos. Among the mere amateurs, sixteen-year-old Molly Nicander from Riverside keeps two snakes, both of which deeply enjoy the company of feeder mice. She looks for "two of the ugliest males around" (females being more valuable for breeding), and then drops them into the snake cage and runs away. Even Debbie Prosser, the sympathetic show judge, keeps a king snake and feeds it what she calls "pinkies." If a breeder of champion golden retrievers were to admit hurling live pups over the fence to her pet tiger, she would doubtless be drummed out of the American Kennel Club. But for mousers, who have come to accept Blitzkrieg reproduction as the rule, a handy snake is simply a way of mimicking the balance of nature.

For people who don't fancy mice, snakes are probably not a very good solution to rampant reproduction, at least not in the house. Indeed, there doesn't seem to be a really good solution, and this begins to suggest the problem people have with mice. Partly it is a matter of cultural or psychological trespass: Mice cross the lawns and fences we diligently maintain as barriers against the natural world. They infiltrate our homes furtively and without invitation. We come to know them only as phantoms—a pattering of footsteps overhead (wild mice given exercise wheels can achieve sixty thousand revolutions in a single night), a scattering of droppings under the kitchen sink (a single mouse voids fifty or more droppings daily and can spread salmonellosis), a blur of movement just off to one side of the hearth (that kinesthetic sense means mice know nooks and crannies of the house that we can hardly imagine, and it also means they can vanish before the hurled book has come to earth or the word "mouse" has been uttered). We begin to get an unsettling sense that mice have made *us* their dominion.

And it's never just one mouse. No, the reproductive possibilities are nightmarish. Consider the phenomenon of mouse plagues. For reasons no one understands, mice have at times descended on communities, or welled up from within communities, in truly horrific numbers. Witnesses to a spectacular Australian plague in 1917 reported that 544 *tons* of house mice were caught in just one town. In Kern County, California, in 1926, an estimated four billion mice—literally ankle-deep waves of them—infested an area eighteen miles in diameter over four months, far too many mice for even a very playful cat.

But since there seems to be precious little we can do about mice, perhaps we should try to think of them finally in a more benign context. The truth is that mouse plagues are rare. And considering their ubiquity and the potential for destruction, mice really do relatively little damage. They offend our sensibilities more than our pocketbooks.

If one searches high and low, and in and out of years, it is even possible to find an instance on record of mice actually benefiting humans. Each autumn in eastern Europe, so-called spike, or mound, mice, which are a species closely related to house mice, prepare for winter by constructing mounds of earth in the fields. Each mound has a core of pure grain, up to fifteen pounds of corn, millet, or barley gleanings left behind by farmers. The mice are assiduous not only about gathering grain, but about protecting it. They steadily repair the mound, which also contains their nest and two or three labyrinthine entry passages, until the first snowfall socks them in.

It has occurred to farmers that these grain hoards might feed livestock as well as mice, and so they raid the mouse pantry. "Ukrainian farmers do not shy from driving their wagons into the fields in the fall to collect the contents of the hills," wrote Austrian researcher Antal Festetics, in 1961. The farmers have doubtless inured themselves to wreaking havoc on mice. They are not known to write poems of regret or to utter the Ukrainian equivalent of Burns's lament that the evicted mice will have "to thole the winter's sleety dribble, An' cranreuch cauld!" Still, it is possible to infer that they are not completely heartless toward their rodent benefactors. "It is said," Festetics wrote, "that if the winter storages of the mice are destroyed early enough, they will build new ones." Alas, it remains unclear whether the purpose of early destruction is to keep the mice alive through the winter for next year's gleaning, or merely to accomplish a double gleaning of this year's crop, and to hell with the mice.

If we can suppress the primordial urge to cry "eek!" a moment longer, it is even arguable that the house mouse has done good—or, anyway, not much harm—in a larger ecological sense. First, it seems to have conquered the world without causing the extinction of any other species. Unlike the notorious rabbit, for instance, it does not bully its competitors out of their niches or wipe out their food supply with its voracious nibbling. It has merely been adept at filling vacancies—most often vacancies that humans have created. Second, the house mouse has unintentionally provided food not just for Ukrainian livestock, but for innumerable other creatures, notably snakes, weasels, skunks, foxes, hawks, crows, barn owls, and house cats, all of whom doubtless agree that mice are nice.

Let us return then to our original question: Do we applaud the house mouse? Taken individually, it is, as the naturalist Georges-Louis Buffon wrote, "a beautiful creature; its skin sleek and soft; its eyes bright and lively; all its limbs . . . formed with exquisite delicacy." And it is also, as Burns wrote, "our poor, earth-born companion," its life inextricably tied up with ours, equaling our success as a species, but not our destructiveness. So do we clap?

I say we give the house mouse a standing ovation—and just for a moment suppress the endless, irresistible urge to stomp the little buggers out underfoot, once and for all.

Cormorant
Heaven

On a winter afternoon in the flat, wet landscape of the Mississippi
Delta, it is a common sight these days to see an untidy string of ten
or twenty cormorants thrashing across the gray horizon. As they
head west, the cormorants link up with another string coming in
from their feeding grounds, and then another, until, as they pass
overhead, they form a long, undulating strand of perhaps two
hundred dusky birds. At their roost, which may be thirty miles
away, as many as five thousand cormorants will fly in for the
night. In the bare branches of a tupelo-and-cypress swamp, they
perch fifty to sixty in every tree for hundreds of yards around, and
they keep coming until dark, as if a painter were spattering India
ink over and over across the treetops.

In the Delta, as in much of North America, cormorants were
uncommon as little as ten years ago. But in the early 1970s, the
bird began an extraordinary recovery from the effects of pesti-
cides and human persecution, and it is now rapidly reclaiming its
old range. This has caused little joy and lots of hostile muttering
from San Diego to the Chesapeake Bay.

Human antipathy toward cormorants dates back at least to Deuteronomy, which declared them unclean, and this ancient, almost pathological, ill will has revived among fishermen and fish farmers everywhere, but especially in the Mississippi Delta. Progressive farmers there began kissing cotton good-bye in the late 1970s, scooping out the local buckshot clay to create 110,000 acres of high-profit catfish ponds. On their annual migration from the Great Lakes to the Gulf Coast, cormorants spotted these rich new wetlands and promptly reappraised their vacation plans. A new destination was born: The Delta had become the catfish-farming capital of the world just in time to become the cormorant capital, too.

On any winter evening, forty thousand cormorants now return to roosts around the Delta. Traveling with them, in cargo, are perhaps thirty thousand pounds of catfish, which might be worth $23,000 wholesale were they not already in advanced stages of digestion. It is a thought to stir feelings of profound theological turmoil in a catfish farmer's belly. "It's horrible," says one. "There ain't nothing good about 'em. I'm sure God put 'em here for something, but I don't know what it is."

Shoot. What any former cotton farmer knows in his heart is that God put cormorants here to take the place of the boll weevil.

The cormorant, for those of you who haven't had the pleasure, is a large black waterbird resembling a cross between a crow and an eel. Its name literally means "crow of the sea," and it is a member of the pelican order. But while a pelican flies with a kind of gliding grace, neck folded back, the cormorant is all get-ahead motion. It hurtles along with neck out, head slightly up, and wings constantly flapping, clocking fifty miles an hour at times, but somehow looking as if it must struggle just to stay airborne.

The cormorant moves more gracefully underwater than in the air, which is the main reason people in the Delta don't much like it. A catfish farmer will be out tending one of his ponds when a flock of cormorants will splash down two ponds over and float for a few minutes. As the mood arises, the birds lift their wings to

shake the air out of their feathers. Then, one by one, they disappear underwater, apparently using the small round stones often found in their gut as a kind of diver's weight belt. The birds' feathers are interwoven much more loosely than in other waterfowl, making the cormorant more "wettable" and less buoyant. It swims underwater with its wings folded along its slender body, its long, sinuous neck curving inquisitively from side to side, and its large eyes alert behind clear inner lids, like the nictitating membranes in certain reptiles. A swimming cormorant is a bit like Daffy Duck in goggles.

Simultaneous thrusts of its webbed feet provide enough propulsion for a cormorant to tailgate a fish and catch it crosswise in its hooked bill. It can catch fish even at a depth of eighty feet, staying under for as long as a minute. By comparison, a short dip in a six-foot-deep pond boiling with, say, 175,000 catfish is as easy as Chinese takeout. The cormorant generally brings a fish to the surface after ten to twenty seconds and flips it in the air to position it correctly and smooth down its spines. Sometimes a flock of cormorants will line up and herd the catfish into a corner. Then they take wing again with full gullets and move two or three ponds farther along just about the time the farmhands arrive at the first pond with their shotguns, pyrotechnic "bird bangers" and all the other futile instruments of anticormorant outrage.

Even without this catfish cafeteria, the cormorant is wily enough to get by on about a half-hour of honest work per day, if one can properly apply that term to what is, after all, fishing. Otherwise, cormorants mostly loaf. You can see them hanging around on rivers, lakes, and harbors almost anywhere in the world. Their appearance always suggests something other than cormorants. Hence such common nicknames as "water turkey," "crow duck," and even "shark."

On the water, they ride low, with only their necks and heads visible, like submarines with their periscopes up or miniature Loch Ness monsters. On rocks or pilings, they like to perch upright with their scraggly wings spread out to dry, Dracula-fashion. In breeding season, many of the twenty-five or more cormorant species sprout feathery crests. The most common of these in North America is known as the double-crested cormorant, though for much of the year its head is as smooth and slicked-back as a tango dancer's. This combination of mutability with a netherworldly appearance has sometimes evoked images of the devil. In *Paradise Lost,* for example, John Milton wrote that Satan "sat like a cormorant" on the Tree of Life, preparing to work mischief in the Garden of Eden. It is almost enough to make the catfish farmers of Mississippi attempt exorcism.

They have tried everything else. A couple of seasons back, for example, a radio-controlled-aircraft enthusiast named Cooper Kimbrell spent several mornings waging a small-scale Battle of Britain over a catfish farm. Kimbrell is happy to re-create that moment in history for a visitor. He stands with the tail of his aircraft restrained between his ankles as the engine whines up to full speed. On the fuselage, a painted "storm lizard" extends a forked tongue and appears to flex its claws cormorantward. When Kimbrell releases the plane, it climbs rapidly, and he puts it through loops and snap rolls, and then sends it screaming down like a dive-bomber. "A bird hears that coming at him," he says, "he turns and runs the other way."

To try out the technique for local fish farmers, Kimbrell and a couple of fellow hobbyists positioned themselves on the dirt levees dividing the catfish ponds and waited for a squadron of cormorants to approach from its night roost. When birds began to peel off for a descent, Kimbrell's plane went up to meet them: "I pointed it straight up and held it there and went straight up through them and just hoped that I was the one that came out on top." As the cormorants broke right and left, he pursued them, and here Kimbrell sounds like a Spitfire pilot recounting a dogfight: "You had to concentrate at things you weren't sure you could do, and you just did 'em. You found you were a better instinctive pilot." Kimbrell stayed on one bird's tail close enough that the cormorant craned its head back in mid-flight to see what was chasing it, then folded its wings and plunged underwater from fifteen feet up. The catfish farmers thought it was the best show they had seen in years. In the cold light of dawn they also realized that aerial warfare would cost money, not just for wages, but for the planes, which one agricultural extension agent describes as "expensive little buggers." In a Louisiana trial a plane hit a cormorant, more or less accidentally. The bird lived. The plane didn't.

Most farms now rely on less entertaining methods. Butane cannons boom at intervals across the ponds—and also provide perches for cormorants, which are smart enough to know they are firing blanks. On some farms a jack-in-the-box scarecrow, called "man-in-a-cannon," leaps out of a fifty-five-gallon drum with an ineffectual bang. Other farms have experimented with a device called "air crow," which deploys revolving strobe lights and inflatable plastic arms, along with the sound of horns honking, people shouting, shotguns firing, and birds screaming at a reported volume of 130 decibels. But a cormorant can get used to almost anything.

Killing them has of course occurred to one or two fish farmers in bleak moments of aquacultural despair. But cormorants have been afforded federal protection under the Migratory Bird Treaty Act since 1972. In some cases, the U.S. Fish and Wildlife Service (USFWS) will issue a permit allowing a farmer to kill up to two hundred or so birds a year, with the idea of making the air crows and other instruments of harassment more credible to discerning cormorants. But the paperwork is onerous, and farmers protest—perhaps too much—that killing cormorants isn't easy. They fly away if people get close, and at a distance, especially on the water surface, they make meager targets. "They're tough birds," a farmer says. "Nothing to 'em but bones. There's nothing to kill."

That is not quite the case. In California, the USFWS convicted one fish farm with a fifty-bird permit of illegally killing fifteen thousand birds over five years, including herons as well as cormorants. Minimum-wage field hands, ill equipped to distinguish between a water turkey and a bald eagle, apparently blazed away at anything with wings, even hawks and ducks that don't consume fish.

The result is that battle lines are beginning to form. With aquaculture expanding and natural wetlands dwindling because of development, fish farms can become a magnet for an area's entire population of some waterfowl species. Because of the risk to these birds, federal officials are reluctant to issue special cormorant-killing permits. Angry fish farmers, who are by and large wealthy businesspeople, have turned to Congress for help. One result is that researchers from the U.S. Department of Agriculture's controversial Animal Damage Control unit, best known for killing coyotes to benefit western ranchers, have recently moved into the Delta to contemplate the cormorant.

Fish farmers don't talk about what they hope will result, except that it should be "something big-time, in the roost, with federal sanction." Privately they are pushing for a standard depredation order, which would allow liberalized killing of cormorants even on open waters. To make this politically more palatable, they are trying to convince people that the country is being overrun by cormorants. This is, of course, merely God's honest truth to a catfish farmer who did not even see his first cormorant until seven years ago but recently had sixteen thousand of them roosting in sight of his ponds, or to one Texas shrimp farmer whose cormorant losses were estimated at $25,000 a week "till he drained the dagblamed ponds." They see themselves under siege. "Y'all give us a break," an equipment supplier calls out when he spots a reporter doing interviews at a catfish restaurant in Belzoni, Mississippi. "We ain't trying to destroy them. They're trying to destroy us."

This is a venerable sentiment among fishermen, who have fumed for centuries over the cormorant's ability to go underwater and catch fish when intelligent humans can't even get a nibble. Idle minds have often leapt to the conclusion that the cormorants were taking all the fish. From there it was another small leap to the misguided idea of conserving game by killing cormorants. As in the case of bats, killing cormorants was particularly easy because of ingrained human prejudice against them.

Not only does the cormorant resemble the devil, but its gullet is large and its purported greed is legendary. Though the typical adult stands just three feet tall, cormorants have been known to take fish of up to sixteen inches in length or to contain as many as eight smaller fish at a time, in various stages of digestion. Chaucer considered them synonymous with gluttony, and Shakespeare described Time, that devourer of life, as a cormorant. Usurers were once "money-cormorants," and the bird itself has in places been nicknamed "lawyer," presumably because of its alleged avarice. These unsavory cultural associations persisted into the 1970s, when a faction of Cambodia's genocidal Khmer Rouge revolutionaries became known as "cormorants."

In China and Japan, the historical response of fishermen to the apparent competition from cormorants was to domesticate them and work them like prized retrievers, using a hemp or leather collar to keep the birds from swallowing their catch. The usual Western response, on the other hand, was mayhem. Queen Elizabeth I authorized a bounty on cormorants and other "ravening birds and vermin." In the United States, fishermen placed leg-hold traps on perching sites and visited nesting colonies in the summer to drop large rocks on the eggs. On top of this persecution, widespread use of the pesticide DDT caused cormorant eggshells to become so thin that they crumbled under the brooding parent's own feet. In Wisconsin, where huge flocks once nested, only sixty-six breeding pairs remained by 1972.

The most peculiar thing about the long persecution of cormorants is that there appears to have been hardly any reason for it. Over the past century, more than 125 scientific papers have examined the cormorant's dietary predilections. They indicate that in normal circumstances, the cormorant mainly catches fish in which sport fishermen have no interest—alewives, for example, in the Great Lakes. It also preys on suckers, sticklebacks, ling, sculpin, and other species that eat the spawn of commercially valuable fish. So in theory, the cormorant's choice of food may actually benefit humans. But sportsmen tend to believe what they think they see with their own eyes. For example, near Watertown, New York, on Lake Ontario, more than eight thousand cormorant breeding pairs now nest on Little Galloo Island. Local fishing guides have accused the birds of eating trout, bass, and salmon fingerlings stocked by the state. A simple solution proposed by biologists was to stock fingerlings at night, when cormorants don't fish, to give them a chance to disperse. But even this proved unnecessary on more careful consideration: The fish found liberally strewn around nests on Little Galloo almost always turned out to be alewives, not salmon or trout. The guides have nonetheless petitioned for an open season on cormorants, so far unsuccessfully. Then, in the summer of 1998, unknown vigilantes went ashore on Little Galloo with shotguns and slaughtered almost 1,000 cormorants, most of them fledglings still in the nest. Federal officials called it an act of "environmental terrorism." But local fishermen likened the incident to the Boston Tea Party, as a blow for their freedom.

The one place where humans have a legitimate gripe against cormorants is where we have ourselves created an unnatural concentration of fish. Cormorants, always opportunists, are old hands at this sort of thing. In the J. Paul Getty Museum in Malibu, California, a painting by Vittore Carpaccio from 1495 depicts fishing enclosures on a Venetian lagoon. Cormorants are everywhere—ducking underwater, floating with only their heads and necks visible, perching on the fishermen's fences. Seven boats patrol the waters, and in the bow of each boat an archer fires pellets at the birds. What seems to be going on here is a fifteenth-century version of predator control—and for the same reason as in the Mississippi Delta: Cormorants can pose a real threat to any concentrated fishery. Federal researchers have examined the stomach contents of 136 cormorants in the fish farms of the Mississippi Delta and found that 64 percent of their diet consisted of catfish. The fish farmers themselves estimated their direct losses at $3 million a year, plus $2.1 million for butane cannons and the like.

Losses from other factors can be at least as bad. Indeed according to the U.S. Department of Agriculture's own statistics, 70 percent of catfish losses nationally are due to disease and another 10 percent to oxygen depletion; birds account for only 7 percent of the total. But none of these threats is so damnably visible as the cormorant, nor so well suited for an angry letter to one's congressperson. Moreover, there is something inwardly rankling about the knowledge that the cormorant problem is getting worse at least in part because of the fish farmers themselves. A full winter of rich southern food is surely one reason why cormorants nesting in the Great Lakes are now showing population increases of 15 to 63 percent a year. "They have a cafeteria here," a spokesman for the Delta catfish farmers concedes. "They head North with a half-inch of fat on them, and all four eggs hatch, and then—here they come again."

• • •

On Little Gull Island at the mouth of Wisconsin's vast Green Bay, seven hundred pairs of cormorants now nest each summer. The nests are whitened platforms of interwoven branches, clustered together everywhere on the rocks and in the trees. Having been reoccupied and continually improved over the years, some stand a foot tall, or more. The males move in first each spring and sing seductively to every passing female. Researchers have unkindly likened the cormorant's *oak-oak-oak* love song to the sound of human belching or to "the series of choking sounds sometimes made by a cat, but . . . more rapid." Cormorants nonetheless have twice the breeding rate of other waterfowl. Moreover, they form affectionate couples, with the male frequently bringing the female gifts of fresh nesting material, and with both parents sharing chick-rearing duties. Human or other intrusion causes them to flee the area temporarily, leaving their young vulnerable to heat, vandalism, or predation by the herring gulls that commonly share the nesting site. Otherwise, they are devoted and attentive parents.

When people try to describe cormorants, the words "primitive" or "reptilian" often come to mind, and a cormorant nesting site looks much as one would imagine a dinosaur breeding colony. Guano has whitewashed the rocks, and the phosphoric acid in it has killed the vegetation and turned the trees where some birds perch into ruined hulks. Cormorants old enough to leave the nest stand together in large groups, each of them like an upright, elongated "S" written in heavy ink across a bleached-out wasteland. The short, underdeveloped-looking wings and the shape of the skull suggest *Archaeopteryx,* the putative link between dinosaurs and birds. To pull itself up out of a crevice, a cormorant uses its wings, its scaly reptilian feet, and finally, its hooked bill to find a purchase and clamber up onto the rock surface. It is as if evolution were occurring in a single animal's lifetime, or as if the cormorant were a creature in metamorphosis.

Younger birds still confined to the nest rear up when a human approaches and squeal *aaah-aaah-aaah,* with their pinkish yellow mouths gaping in fear. Gular sacs, fleshy evaporative cooling devices at the top of the throat, flutter, and their heads waggle anxiously as if to say, "No-no-no, not me, not me." Several nests contain featherless hatchlings only a few days old; the bones move visibly beneath their leathery black skin.

The human intruder today is a population ecologist named Jim Ludwig, who is president of the Michigan Audubon Society and also of his own firm, Ecological Research Services, in Ann Arbor. Ludwig is a burly fiftyish man with a jutting jaw and blue eyes behind thick, smoke-colored bifocals. He wears a T-shirt and stained blue jeans, and his forearms are so cut up from handling cormorants that when he recently tried to obtain hypodermic needles to take blood samples, the nurse mistook him for a junkie. He started studying birds on Lakes Huron and Michigan in 1959, and for the first ten years he did not see a single nesting cormorant. By the early 1990s he was visiting twenty-eight cormorant nesting colonies in the same territory each summer, logging seventy-five hundred miles in his battered research boat.

On Little Gull, Ludwig approaches a group of five- and six-week-old birds, and then suddenly bounds in among them, grabbing four by the neck and carrying them out, limp as dead turkeys. He sits down on the rocks and tucks the birds under his legs, talking to them gently. Then he wraps a towel around one bird's head, stretches out its wings, and lays it on its back across his lap so an assistant can take a blood sample. Afterward, the bird shakes itself and stumbles off even more awkwardly than usual.

While the cormorant's reproductive prowess may be daunting for fish farmers, Ludwig believes it makes the species an ideal early-warning system for environmental hazards, a kind of canary-in-the-mineshaft for the entire Great Lakes. What Ludwig and a small group of Canadian and American researchers have inadvertently been developing is a perverse answer to the Mississippi catfish farmer's question about "what God put 'em here for," or to its inevitable secular corollary, "What use are they to humans, anyway?" One tentative answer is that research on the booming cormorant population may prevent human suffering—particularly among people who eat fish—and spare us from human stupidity. Unfortunately, if the research findings turn out to be correct, they could also produce an economic impact that will make the catfish farmers' woes dwindle into nothingness.

Standing on Little Gull Island, Ludwig likens the Great Lakes, which contain 20 percent of the world's fresh water, to a huge bathtub with no outlet. It takes Green Bay alone fifteen years to achieve a single change of water, he says. What goes into the water stays there, accumulating in the flesh of fish and ultimately in the people or cormorants that eat the fish. What went into the Great Lakes in the 1960s and 1970s were polychlorinated biphenyls, or PCBs, organic compounds used in lubricants, heat-transfer fluids, and about three hundred other products. Most government monitoring programs now show the total PCB burden in the Great Lakes to be declining as a result of a 1976 ban on such compounds, which were found to be carcinogenic. But Ludwig calls the reassurance false. "We believe U.S. state and federal regulatory agencies have measured generally the wrong things with the wrong techniques, and failed to look at the real world to ascertain if fewer biological effects have occurred in step with reduced contamination as measured by totals."

Waterbirds can concentrate toxic residues at twenty-five million times the levels present in the water, and Ludwig routinely finds PCBs in his cormorant blood samples. Moreover, he believes that the decline in the total PCB level in the lakes may disguise increased proportions of the PCBs that are most toxic for cormorants and humans. The booming cormorant population provides a means of quantifying the effects of this change, and what Ludwig is learning is that reproductive defects and deformities are increasing. Crouching by the edge of the water with a cartonful of dead cormorant eggs, Ludwig cracks them one by one and examines their contents. Most are infertile, but after three eggs he cradles an underdeveloped embryo in his fingers: "Twenty-five days," he says. "Severe head and neck edema. Let's jar this one. This is one of the PCB- or dioxin-like symptoms we're seeing." In three of the dozen eggs, the embryos have these swollen, goiterous necks. Ludwig notes that detecting such a phenomenon in mammals would be difficult, since defective embryos tend to be absorbed into the lining of the uterus or aborted and lost. Because cormorants develop so rapidly after hatching, they also display growth abnormalities more noticeably than do mammals. On a single visit to Little Gull in 1990, Ludwig found seven hatchlings with crossed bills, four with eye defects, and a couple with leg deformities.

Other researchers from the United States and Canada have corroborated Ludwig's findings of reproductive impairments in cormorants and other species, including bald eagles and trout. One survey of mammals that eat only fish suggests that some mink and otters have stopped reproducing altogether along much of the Great Lakes shoreline. Other studies in humans show that women who have consumed PCB-laden fish from Lake Michigan give birth to children with smaller head size and poorer neuro-muscular development. Public health departments advise against eating Great Lakes fish regularly, if at all. But no government agency has yet followed Ludwig in making the leap from cor-morants to humans and concluding that reproductive problems are getting worse because of PCBs. "The implication of declaring that this is what's going on," says one scientist, would be to neces-sitate "a multibillion-dollar cleanup."

It would be a lovely, if bittersweet, turnabout were such a cleanup to become a moral necessity because of what researchers can see happening in cormorants. The bird might then be trans-formed in the convoluted human mind from agent of the devil to environmental archangel. But this is surely wishful thinking. A bird that enlightens society about a problem—one for which the enormous costs of a solution are far easier to calculate than the eventual savings in human health—is always going to seem more like a headache than a hero.

What we can extract from the cormorant's strange tale, then, is merely a lesson in how complex nature can be, how difficult it is to characterize its changing manifestations as "good" or "bad," and how cautious we should be about the urge to tamper with it. There are unforeseen consequences in everything we do.

For the catfish farmers, who are in any case unlikely to win approval to kill large numbers of cormorants, it may still be possible to obtain a measure of peace. Rather than beseeching Uncle Sam for relief, it may prove less troublesome in the end for them to sit back and shrewdly consider their adversary, and then adapt their own methods accordingly. Cormorants are not great aerialists, for example. They need room to land and even more room to take off. With its wings flapping a cormorant may run along the surface of a catfish pond for thirty to forty yards, splashing like a kid in loose galoshes, before it becomes airborne. Putting a net across a pond to keep cormorants out costs too much—$22,000 for a twenty-acre pond, in one estimate. So researchers at Texas A & M University dug long, narrow ponds, and then strung black twine across them at thirty-foot intervals, perpendicular to the prevailing winds. For a total cost of $13 an acre, they found they could play havoc with cormorant flight patterns and save catfish for harvest.

This elegant solution may not come as great news to a farmer who has just invested huge sums to dig five hundred acres worth of big blocky catfish ponds laid out like cormorant runways. But there is another alternative: He can work on improving his harassment methods and learn to swallow his cormorant losses as just another cost of doing business, like disease or oxygen depletion. Relief may not come immediately, but it will come. Eagles, owls, or some other predators will move in on the cormorants—or the alewife population in the Great Lakes will nosedive. PCBs may ultimately overcome the cormorant's reproductive capabilities, as did DDT. One way or another, the booming cormorant population will go bust. Unlike the devil, cormorants are not eternal. Or as a wildlife agent puts it, a tad ruefully, "Didn't there used to be something called the balance of nature?"

Acting Like Animals

It is 8:00 A.M. on a hot January day in Hollywood. On stage five at Raleigh Studios, a film crew is adjusting lights, checking cameras, rigging props. The star of the commercial to be filmed today stands in a darkened corner, naked except for the grip's tape wrapped protectively around his feet. In person he is surprisingly short, like so many film stars; counting the tape, he stands just fifty-eight inches tall. His main lady has stayed at home this morning with their firstborn, but he is accompanied by a soothing stand-in girlfriend.

Merrill (a star in Hollywood needs no last name) is also surrounded by a small army of retainers. Four of them are now spraying him with a brown vegetable-based hair coloring called Streaks 'n Tips. He will require 120 cans of the stuff over the course of the day. Merrill, whose natural coloring is white with auburn spots, endures the tedious business of makeup about as well as any veteran actor. He has been through it all before, playing the part for which he was born: longhorn bull. He helped win a Clio (the advertising Oscar) with his performance as the bull in the china shop. Today he will play the role of the bull on the chessboard, masterminding the Sicilian defense. He is a pro.

An executive from Merrill Lynch, the brokerage house that is paying for today's shooting, studies the clouds of spray forming a brown halo around the firm's namesake and corporate symbol. "In the all-American tradition, the star's hair is dyed," he says, to put the scene in context. He thinks about the rightness of that for a moment, and then adds, "At least he isn't in drag."

In the world of animals who act, almost anything is possible: a pony that will pose as a green unicorn, a cockatoo that will imitate a windshield wiper, even a dog (if not a fifteen-hundred-pound longhorn bull) in drag.

Show business trainers say they will never take a job that pains or endangers an animal (aesthetic pain is, of course, a separate issue). They also cannot overcome anatomical limitations. Trainer Hubert Wells once got a script in which a seal was supposed to stand on its hind flippers and stroll down the sidewalk eating an ice-cream cone, a notion roughly like asking a ballerina to dance *Swan Lake* on her tongue. Otherwise, for every imaginable role, for any stunt, there is a suitable animal—and probably also an understudy and several rivals—all within an hour's drive of Hollywood. The services of this diverse community are regularly in demand, for a reason both Hollywood and Madison Avenue understand in their souls: Animal actors, running the emotional gamut from A (for Arnold the pig) to B (for Benji), have always been top dog at the box office.

"Is the bull ready?" says a voice from behind the camera.

Tom Mitchell, Merrill's main handler, nods and leads the animal by a halter out onto the enormous chessboard that is the set

for today's commercial. Merrill promptly takes up his first position, off-center on the black king's side of the board. Mitchell
eases the halter off. He and Danny, another wrangler, loop a piece
of hundred-pound-test monofilament that looks like dental floss
around each horn to steady the bull as they slowly back off the
chessboard. The camera starts to whir, and Merrill picks up his
head, instinctively working to it. "We're rolling now," says the
voice. The wranglers nod and draw the monofilament off the bull
and out of the picture. "Okay," someone announces sternly, "the
bull is loose."

The most basic skill for a Hollywood trainer is the ability to read
an animal's mood and, with luck, do something about it before
anybody gets maimed. The task is more difficult than it might
seem, because the animals must routinely share the set with
human actors who know little about them, and who may like
them even less. "Never work with kids and animals," W. C. Fields

Merrill as
Grand Master

is reputed to have said. "They'll steal your best scene with their behinds to the camera."

The task is further complicated because animal actors, like human ones, sometimes learn to disconnect outward behavior from real gut feelings. Hubert Wells has trained wolves to snarl on command and, in the next instant, to nuzzle him for their food reward—an extremely unnatural transition. Trainer Monty Cox teaches big jungle cats to mute their killer instincts even as they knock him to the ground with claws out and fangs bared. The trick is to recognize when feigned anger becomes genuine and when the tiger starts stalking for real.

Cox, interviewed during a practice session with a sixteen-month-old tiger, says it is mostly a matter of working with an animal often enough to discern anger or fatigue in its body position or the look in its eye. Then, having reassured his small audience, he launches into an extensive round of amputee jokes.

On stage five at Raleigh Studios something is not quite right. Merrill drops his head and steps forward. "Watch it!" someone shouts, but the film crew is already scattering. Merrill comes off the chessboard at a lope and is rolling by the time he passes between two wranglers, who try to lasso him unsuccessfully. The bull heads straight for the chair in which his owner, livestock trainer Joan Edwards, is sitting. She does not so much as miss a pull on her cigarette. Five feet in front of her, he turns, as she knows he will, and heads into the enclosure where his girlfriend is waiting.

It is, say his handlers, just a case of the jitters. The pressure on the set is no less real for an animal actor than it is for a human one, and trainers routinely bring along extra animals as companions to give their working stars something to escape to. (Odd instances of bonding abound, as between Marlon Brando and Wally Cox. Trainer Steve Martin of Working Wildlife has a leopard that is never truly at peace except in the company of the malamute with which it was raised.) In any case, Merrill regains his composure, and filming proceeds in earnest.

Trainers go to extraordinary and imaginative lengths to get the performance they want from an animal. For *Ring of Bright Water,* Hubert Wells spent months training otters, which are about as sedentary as quicksilver, to go to a spot the size of a quarter and stand there for twenty seconds. Since close-ups in a movie rarely last longer than fifteen seconds, this was, he says, just long enough. The usual incentive for the animal is food. But David McKelvey, an animal consultant, sometimes employs his repertoire of about three hundred animal calls as a persuader. To train a rooster, he once imitated a hen in heat.

Instead of teaching the animal a routine, which takes time, trainers sometimes devise a gimmick to trick the animal into doing what the script requires. Not long ago, a beer company wanted Joan Edwards to get a bull to jump through a seemingly solid brick wall—a feat that goes against the animal's instinct for self-preservation. Edwards knew that bulls don't like the dark; they will move toward light. So she had the set designer build the wall with a dark, narrow eighteen-foot runway behind it. The only light came from holes between the artificial bricks. As a finishing touch, Edwards had a white strip painted across the bottom of the back side of the wall; the appearance of a low obstacle would cause the bull to leap. In the resulting commercial, the bull (an aggressive sort named Heckle) practically flies through the wall.

The trainers themselves sometimes become a kind of gimmick. Diane Brisby, of the Exotic Animal Training and Management Program at Moorpark College outside Los Angeles, raised a female cougar from the age of nine days. But when the cougar became an adult, Diane apparently went from mother figure to rival. Now, says Diane, if a scene calls for the cougar to have fangs out and ears back, "All I have to do is say, 'Hi, Eve.' " In the jargon of the trade, she has become Eve's "hate person." Like a court jester who has told the wrong joke, a hate person should not show up again on the set after provoking an animal to snarl. Diane, fingering her throat, says it is all very depressing.

Like that other great slab of beefcake, Sly Stallone, Merrill is not at his finest in speaking parts. What he does best is simply look good, and today's script plays to that talent. Merrill must rouse himself from his normal disposition, which is placid, bordering on torpid, only enough to project something like intelligence. The huge chess pieces around him ride on casters, with monofilament control wires being worked by stagehands off-camera; Merrill will move each piece with a mere glance or turn of his head, as if by telepathy.

The trick is to get the right sequence of head turns and chess moves. At midafternoon, the crew is filming its umpteenth take of the fourth featured move (all moves come from a classic match between the former Yugoslavia and the former USSR).

"Tom, put his ass to the right," says the voice from behind the camera. Mitchell nudges the bull over as the camera gets up to speed. "Now, Danny, get him to turn his head toward you." Danny, off to the right, begins to slap the walls and shout, "Huh! Huh! Merrill! Merrill!" The bull is indifferent. Then Danny beats the floor, and Merrill executes a brilliant head turn.

"Send the piece," says the voice. White queen slides to queen's knight one. Black rook moves in turn to king seven.

"Now get him to look at you, Tommy."

Mitchell, just under the camera, wearing white cotton booties to protect the set, scuttles forward shaking a sheet of brown paper in one hand and rattling a galvanized bucket with a 7-Up can inside in the other. Merrill turns and looks at him. Depending on one's reading of the bull, he is saying either, "Bobby Fischer couldn't have played it better" or "What in the Sam Hill are you doing down there, Tommy Mitchell?"

"Rock 'n' roll!" a stagehand whoops, not pausing to quibble about nuances. Says the voice behind the camera: "That's a cut."

It is sometimes hard to tell if the world of animal actors is a parody or merely the epitome of the splendid follies of Tinseltown. *Item:* Until recently, the animals had their own Patsy Award instead of the Oscar. Frank Inn, a wealthy trainer who kept humble digs opposite D & D Auto Parts and Salvage in Sun Valley, had a wall covered with forty Patsys. The plaques hung just across his dining table from the stacked funeral urns of the animals who won them. *Item:* The ranks of aspiring animal stars are sprinkled with shamelessly familiar names: snakes called Boa Derek and Julius Squeezer, a buffalo named Cicely Bison and a bird named Martha Rhea, even a ferret called Ferret Fawcett-Majors. *Item:* There is no Schwab's drugstore for would-be Lana Turners of the quadruped variety, but Mac, an alley cat featured in *Logan's Run,* was discovered in a garbage bin behind a McDonald's franchise. The trainers of both Sandy (last seen on Broadway in *Annie*) and Morris the Cat (sometimes known as "Mortz" but never, never, not even in Hollywood, as "Mo, baby") say they discovered their stars at death's doorway in the pound. And the original Benji was a worn-out face card from the 1960s, a thirteen-year-old veteran of *Petticoat Junction,* when a producer spotted him and saw box-office gold. The rest is history. *Item:* In case the rest is not *quite* history, the leading animal actors, like their human counterparts, have retainers who are keenly sensitive to the value of publicity. Benji's slick promotional package quotes the *New York Daily News:* "Benji's uncanny ability to project emotions definitely makes him the Laurence Olivier of the dog world." Lassie, the real Olivier, is of course above this sort of thing.

• • •

On the set, the film crew is shooting a brief sequence without Merrill. Tom Mitchell sits in a corner with a device he has manufactured to make rude noises every time the voice behind the camera says "Action." The bull is asleep. Outside, Joan Edwards is giving an interview to a television reporter.

"What's the hardest part for Merrill?" the interviewer asks.

Says Edwards, with a sympathetic nod for her flagging star: "I'd have to say it's the long days."

The world of animal actors is, finally, a business world. For the working stiffs, the money is not spectacular. An ND (or nondescript) horse is $35 a day, a chimp $400, and big cats $500 to $1,000. But an animal who breaks through to stardom can write his own ticket, and that may be why the competition is so fierce.

Benji, billed by his handlers as "the free world's most huggable hero," became an industry unto himself after his first film grossed $45 million. He has starred in three other films and four television specials. His name and image have been sold for T-shirts, sleeping bags, space suits, trash cans, and shoelaces. Updike and Mailer have publishers; Benji has several "publishing licensees to promote his various books and book-related products." He also recently licensed a Benji computer game. The corporation that controls his name calls this "a very exciting license with staggering exposure potential." Finally, there are the personal appearances, or "exposure events," around the nation. Benji keeps up a schedule that is busier than George Will's and commands a speaker's fee approaching that of Henry Kissinger. Mickey Rooney should be so cute.

It is, of course, possible to survive at the top and still keep up standards. Lassie, who is by now a member of Hollywood's Old Guard (the current Lassie is the seventh in the line), manages to combine fabulous wealth with a genial, dignified lifestyle. But lacking Lassie's deeper hold on the public, most of the stars continue to hustle even at the top. Their trainers must constantly aim to do something different, to find some new twist to keep up box-office interest for another year—even if this means that the animal must learn, in effect, to dance *Swan Lake* on its tongue.

Thus a few years ago, Benji's corporation decreed that he should become a scuba diver. A diving suit was manufactured at a cost of $10,000, and Frank Inn began training the dog. They started out with the dog swimming on the surface, to get him used to the suit. Inn talked to him via hydrophone and "reinforced him" with pieces of steak fed through a cork in the dome of the diving suit. Using weights, Inn then got the dog accustomed to working underwater. After several months the dog was ready, and filming began. It is tempting to imagine Benji standing on the bottom of the pool in his diving suit like Benjamin Braddock in *The Graduate,* contemplating the surreal expectations of all those mad people up there above the air bubbles. The problem with this analogy is that Benji was almost certainly glad to fulfill those expectations, because Benji will do anything for a piece of steak. Back home, where Benji has just been induced to climb a ladder backward and dive blindfolded into a pool from a high platform, Frank Inn says it all: "Isn't he amazing? He's never done that before. There's nothing that dog won't do."

It is 11:00 P.M. on stage five at Raleigh Studios, and the crew is trying to get the final move of the game on film. "Can you get his head up, Tommy?" says the voice from behind the camera. "He looks like he lost." Mitchell raises the bull's head. The chess piece rolls. The head turns work. Someone says, "Back to position one, please." But it is a cut, a take. They have it. Checkmate.

The voice from behind the camera asks for a few close-ups—head-on, right profile, left profile—to be edited in as needed.

"O.K.," the voice says finally at 11:21. "Wrap the bull."

Nineteen days later, the finished commercial is broadcast nationwide during the Super Bowl. It runs thirty seconds, and is seen and admired by eighty-two million people. There is talk of another Clio.

On his ranch north of Los Angeles, Merrill spends the day relaxing with friends but does not catch the game.

A Mouse Like
a Spear

One Sunday not long ago, at the height of the English autumn, I went out walking with a researcher named Carolyn King in Wytham Wood, on a hill overlooking the domes and spires of Oxford University. We took a path known as "The Singing Way," once a pilgrim route to the convent of a Saxon holy woman. Ahead was Great Wood, where Oxford field researchers have been studying wildlife ecology for fifty years. Behind was Marley Wood, where King did her doctoral research on the secret lives of weasels.

The scene felt like of one of those classic animal tales of the English countryside. King, a slim, forward-leaning woman in her fifties, with an amiable grin turning up the right corner of her mouth, pointed out the badger paths crossing our trail and the places where she tried to figure out, with the help of live traps, what the weasels were up to as they darted about "like small bolts of brown lightning."

She indicated the rusty bracken on either side of the trail, the rodent holes among the roots, and the gnarled and regal old oaks overhead. "I love the thought of the animals living their private lives under all this stuff, and occasionally we come in and ask questions," she said.

The questions King asked, first at Oxford, and later as a researcher and instructor at Waikato University in New Zealand, had to do with what the weasels were eating, where they traveled, and how they managed to survive in a cold and hostile world despite having short fur, little fat, and the metabolism of a hip-hop dancer on a caffeine bender.

"It's just too bad we have to trap them to answer the questions," said King, who was back at Oxford on a yearlong sabbatical. "I used to fantasize about learning their language and sending a message around with the robins, and they would all gather together and say, 'Look, we've got a few questions to answer.' But then I don't know how you would write a thesis on the basis of that methodology."

The late afternoon sun cast long shadows, and it occurred to me that there was only one problem with the animal-tale character of the day: In most traditional stories, King's heroes, the weasels, would be the archfiends—"bloodthirsty villains" laying waste to "the poor faithful creatures" of Toad Hall in Kenneth Grahame's *The Wind in the Willows,* for instance, or cutthroat soldiers in "the weasel army of Feragho the Assassin" in *Salamandastron,* a novel in the Redwall series by British writer Brian Jacques.

"There is something enormously satisfactory about a weasel," King writes, in her own book, *The Natural History of Weasels and Stoats.* But you would not have a clue what she means from our popular culture. Human references to weasels are commonplace and almost never flattering, not when Washington Irving wrote about a "meagre, weasel-faced Frenchman" nor when David Letterman imagined George Washington's secret opinion that Ben Franklin was "a fat kite-flying weasel."

Weasels are too quick for most people ever to have seen one in the wild, and zoos rarely display them. Yet they have entered our language as emblems of conniving dishonesty (all publishers are "thieving weasels," according to radio personality Don Imus), selfishness (replacement players in the most recent major league baseball strike were "self-deluding weasels taking advantage of another man's suffering," said an ESPN announcer), and cowardice ("gutless weasels" has become a standard epithet, and Calvin, the cartoon hero, once told his friend Hobbes that the middle ground is for "sissy weasels").

The weasel and the stoat (a variety of weasel common in the British Isles) were once associated with the gentry. Their fur, which turns white in winter and is known as ermine, provided the luxurious trim for courtly robes. But even this only brought the weasel death, not glory; King quotes a bit of "anonymous doggerel" suggesting that it was entirely appropriate for the ermine ("rhymes with vermin"), "a bloodthirsty, vicious detestable crook," to adorn "the judge, the dowager, the duke." Nowadays, the ermine trade is much reduced in scale. But "weasel" lives on as a synonym for that subspecies of Hollywood gentry, the entertainment industry executive, conjuring up a whole host of smarmy tassel-shoe traits (as when the *Washington Post* recently wrote about "gutless vacillating network weasels"). Carolyn King calls such slurs "character assassination." When we link weasels with politicians, producers, and the like, she says, we do a terrible injustice to the weasels.

"They are beautiful animals, and I get quietly annoyed when people malign them. But there's no point in saying anything. It's a cultural icon. Human traits get projected onto an animal for no reason, the same way people think of foxes as cunning or owls as wise."

And yet she cannot help pointing out that, far from being gutless, weasels are "bold and confident out of all proportion to their size, and they do not seem to know the rules." According to King, they are one of the few solitary predators capable of taking on prey larger than themselves. A small but zealous weasel once killed seventeen rats in twenty minutes. In another case, a hawk swooped down to seize a weasel in its talons, about as dire a circumstance as a small furry animal can face. And yet as the hawk flapped skyward, its flight suddenly became erratic. A witness watched it tumble from the sky and found the hawk dead on the ground, with the weasel's teeth sunk into its breast. The literature also includes a report of an eagle that managed to avoid being killed by a weasel it had plucked up, but lived thereafter with the weasel's bleached skull permanently latched onto its neck. This is gutlessness only of the most literal, or skeletal, variety. Letterman should be such a sissy.

Not, I admit, that anyone is ever going to make the weasel a hero of song and story. It has too many disconcerting habits—eating bunnies, for instance, and performing the legendary "dance of death" (about which more later). An Irish farmer once told me about being chased down a road late at night by a blood-sucking stoat, which was so intent on his jugular that even after he ran into his house and bolted the door, the stoat scrabbled furiously at the window pane to get at him. ("The Weasel never waits to wonder what it is he's after," the poet Ted Hughes has written, "It's butchery he wants, and BLOOD, and merry belly laughter.") Never mind that weasels are physically incapable of sucking blood. Their mythology is beyond the reach of reason. The same farmer told me he once saw a stoat funeral cortege proceeding solemnly atop a stone wall.

But let's at least begin the process of rehabilitation with that scarcest commodity in the weasel-human dialectic, the facts: There are about ten weasel species worldwide, not including the closely related minks, ferrets, and polecats. Three of them are abundant in the Northern Hemisphere. Both the short-tailed weasel (*Mustela erminea,* also known as the ermine or stoat), and the least weasel (*M. nivalis*) evolved in the Old World and spread to the New by way of the Bering land bridge. (Humans have also introduced both species to New Zealand, where King continued to study them after leaving Oxford in 1972.) The long-tailed weasel (*M. frenata*) evolved in North America and has extended its habitat into South America.

Weasels range in body length from five inches (for the diminutive least weasel) to twelve inches and seldom weigh more than twelve ounces. They're blessed with lithe bodies and short legs, the better to enter rodent burrows; broad flat skulls, to work their powerful jaw muscles on the rodents they find there; and elongated necks, for toting prey without tripping on their own forelegs. Their shape and their darting movements are perfectly expressed in the genus name *Mustela,* from the Latin words *mus* and *telos,* meaning "a mouse like a spear."

Everything about the weasel is geared to the killing of rodents. Despite the weasel's exaggerated reputation for raiding henhouses and bird nests, the weasels in one study directed 99 percent of their attacks against rodents (among them the rats that are the real henhouse villains). This is the job for which they evolved in the first place, King theorizes, roughly four million years ago in Eurasia. By that time, grasslands had replaced forests during a period of global cooling, and early forms of voles, mice, and lemmings proliferated in the new habitat.

Such rodents are the ultimate weasel convenience food, being abundant, nutritious, very nearly defenseless, and, as King puts it, "wrapped in a waterproof package." Predators like owls and foxes are held back from these easy pickings by what she calls "the curtain of grass." They cannot pounce until a rodent surfaces from its burrow. The ancestor of the weasel, some larger forest mustelid like the modern marten, presumably shared this handicap. But natural selection soon favored a smaller, slinkier mustelid capable of following the rodents down into the underworld: the weasel.

Here is what it's like trying to study weasels in the field: I went out one morning with a graduate student working a line of live traps in the English countryside. A rainbow arched overhead, but as we set out, the clouds suddenly darkened and sent a cold, miserable rain pelting down. The graduate student carried a plastic bucket of rotting rabbit guts to smear in front of the traps as a lure. We came back after an hour soaked, stinking, and empty-handed. I thought about King's explanation of why weasel behavior is so little known: "Scientists are human beings, and they like to be sure of getting interesting and reliable results, preferably in the shortest possible time and without getting wet too often." As a rule, weasels don't oblige.

Would-be predators also seem to find weasels somewhat frustrating. Apart from sheer ferocity, weasels possess a battery of defensive mechanisms. They have been known to play dead. If attacked, they can also unleash a stink bomb (the skunk, another mustelid, is a relative). Like their own prey, weasels are generally hidden by the curtain of grass from foxes, owls, and other predators. Even when they hunt in the open, they seem to be acutely aware of the underworld beneath their feet. King once released a weasel and watched it race ten yards straight to the nearest mole hole, as if this escape route were already mapped out in its head.

Weasels also employ camouflage and diversionary tactics for hunting in the open. In northern areas, they turn white in winter to blend in with the snow. Oddly, though, the tails of stoats and long-tailed weasels remain prominently black-tipped. But this, too, turns out to be a defensive mechanism, as demonstrated in an ingenious experiment by Roger Powell, now a zoologist at North Carolina State University. As a graduate student in Chicago, Powell trained hawks to chase model long-tailed weasels wrapped in white fur. Using a pulley, he raced the models across a white-painted roof. The hawks, on loose tethers, attacked the all-white weasels with ease. But when Powell used similar models with black-tipped tails, the hawks hesitated for a crucial moment, as if confused, or directed their attacks at the tail-tip rather than at the vital organs. Those "weasels" escaped, and the black tail-tip was thus established in science as a device to deflect any predator dumb enough to attack a weasel in the first place.

Out in the English countryside the next morning, a hawk soared overhead as I set out with the graduate student in search of

Pop Goes the Weasel

a radio-collared stoat. "It'd be interesting if the hawk came after your stoat," I said, hopefully. The graduate student, who knew exactly how hard it is to get a radio-collar on a stoat, looked at me as if I were mad. "I'd rugby-tackle it," he replied. (Sadly, the graduate student must go nameless here because live-trapping, radio-collaring, and rugby-tackling can all rouse the ire of England's animal rights activists.) For the next two hours, he swept his antenna across fields, meadows, and drainage ditches without success. Just as we were giving up, he cried out, "I've got it," then passed me the earphones to hear the steady signal of the stoat. She was resting out of sight in a scrubby patch of thistle.

We sat down to wait, and the graduate student allowed that he liked weasels at least partly because of their bad reputation. Other predators such as badgers and otters, once also regarded as vermin, are now sanctified in children's literature. But "stoats and weasels are still the little skinheads, the little lager-lout thugs." He paused. "They're very charismatic creatures. Or maybe the word is enigmatic. When you look at a weasel, what strikes you is how intelligent they are, really looking at you carefully. They're very curious, investigative creatures."

Suddenly, the stoat appeared, bounding zigzag across the grass to a drainage ditch. We held back a moment to keep out of sight, and then followed her 150 yards to a meadow, where she immediately began to hunt. Her black-tipped tail flicked from one spot to another in the ankle-high grass. The radio signal went dead one moment, as the stoat disappeared down a burrow, then blipped back again a moment later as she resurfaced. She stood on her hind legs, brown and sleek, showing her creamy white underbelly. Her long, snakelike head pivoted from side to side, frantically scanning the grass to see if she'd stirred up any movement. She leaned forward, as if about to dart at something, then leaped right, and left, and disappeared again, popping up a moment later ten yards away. The grass was spangled with dew, and cows breathed ponderous clouds of steam around their grass-hung mouths. Meanwhile, the mad dog of tunnel warfare was crying "havoc" underfoot.

"I bet she's scaring the shit out of the voles," the graduate student whispered.

There was nothing remotely stealthy about her technique. She was both manic and systematic, demented and yet thorough, qualities that have also been ascribed to the reputed "dance of death," in which the weasel spins and somersaults like a lunatic, causing birds to gawk and draw close. At the height of the dance, the weasel, suddenly sane, darts out and puts a death lock on the neck of the nearest member of the audience. Stoats are also believed to mesmerize rabbits into submission with their musky scent. The stoat may not have enough strength in its upper jaw to penetrate the rabbit's skull and meet the lower canines coming up from the throat. But the rabbit, squealing pitifully, usually dies anyway, of fright. When we left, the stoat was still frantically hunting, and the hawk was wisely looking elsewhere.

. . .

Out at Wytham Wood, Carolyn King was telling me about a return visit to Oxford, after being away in New Zealand for twenty years. At the door to the zoology building, a graduate student asked her to identify herself, and on hearing King's name, exclaimed, "Oh, I know you." For a moment, King enjoyed the warm sensation of having arrived back among the cognoscenti, where her hard work as a field biologist still had its best and truest audience of informed admirers. Then the graduate student added, "You're the one who persuaded the university to pay for a pony as your research vehicle."

King had, in fact, entered university legend as a young graduate student by offering to give up her share in the zoology department's Land Rover in exchange for a pony to run her weasel trapline. "You could buy it for a hundred pounds, run it on grass, and sell it after three years for the same hundred pounds," she argued, with irresistible logic. "Tell me a Land Rover you can do that with." Thus Oxford University acquired an old riding school pony named Jenny, which carried King out every morning in Wytham Wood to check her two-and-a-half-mile-long line of live traps.

"As well as having a companion I could talk to, which was very important, Jenny was helpful," King recalled. "I had to anesthetize the weasels I caught, and vaporization of the ether was very much slower in cold weather. So I carried the ether in Jenny's saddlebag, which had the advantage of keeping it warm. I remember one terribly raw day, my fingers were too cold to move, and I actually couldn't write. I had to go warm my hands under Jenny's mane. Now, O.K., a Land Rover you might have heat in, but Jenny was companionable. She never minded if I warmed my hands under her mane. She was a very lovely animal." The two of them out on "The Singing Way," under a huge old chestnut in the morning sun, were surely one of the most lyrical sights in all of weasel research.

King anesthetized any weasel she caught so she could identify it, take its weight, collect fleas and scats, and send it off again two minutes later, all without being bitten. Considering the weasel's strategy of latching onto victims unto death and beyond, anesthesia seemed prudent. King told me about one dubious theory for getting a biting weasel to let go of your finger: "You hold up another finger and wave it in front of the animal's face. Then, when it lets go of one finger to grab the other, you pull both fingers back. Quickly. I never tested it. I was very lucky never to be bitten by a weasel. But I was very careful. I had huge respect for them."

King was also lucky enough to conduct her study during a great boom in the local weasel population, and in the end she had thirty-seven weasels tagged, most with multiple recaptures. She baited her traps with white mice, which the weasels came to prize. She sometimes released a weasel from one of the first traps of the day only to have it turn up a half-hour later in a trap near the end of the line, perhaps four hundred yards away, "a long way for a tiny little clockwork animal with legs this long."

From her weighing and sampling, King came to see that weasels are not, in truth, little ghouls, any more than they are cowardly or conniving. They are merely carnivores struggling to get by in a world that is at least as perilous for them as for their victims. For example, King suggests that the manic leaping of stoats and weasels might not be a cunning "dance of death," after all. Dissections of weasels commonly reveal masses of bright red parasitic worms packed into tiny cavities within the bone behind the eyes, causing pressure on the brain. King wonders if the dance might not be merely a response to an intensely annoying parasite? Other hazards abound. When a weasel's victim is a rabbit, it may outweigh the weasel by two or three times. A rabbit that has not been "stoated," meaning terrified or mesmerized into immobility, can kick a weasel to death. Indeed, even the hunting of mice and voles can be hazardous; their bites can incapacitate a weasel for days.

The weasel's evolutionary forebears made enormous trade-offs in downsizing to hunt rodents. "The smaller you are, the shorter your lifespan," says King. "All mammals large and small average roughly the same number of heartbeats and breaths in their lives—and these processes are much faster in smaller animals. I once held an anesthetized weasel in my hand, and its heart was running like a little sewing machine . . . *dididididid*." The pulse of a weasel at rest gallops at 400 to 500 beats a minute, King said, a stoat's at just under 400. To supply this hopped-up metabolism, the weasel must eat five or ten meals a day, totaling as much as 35 percent of its body weight.

It cannot afford to miss many meals, because its subterranean hunting grounds make it impractical for the weasel to carry stored energy in the form of body fat. But it also cannot afford to spend a lot of time wandering around looking for something to eat: The small, svelte body shape that is perfect for entering rodent burrows is also the worst possible design for retaining body heat. Unlike most small mammals, the stretched-out weasel cannot even ball itself up to conserve warmth, and it must keep its fur coat thin even in the dead of winter. Moreover, the weasel has no snug little home of its own in which to escape what King calls "the infinite heat sink of the clear night sky." Finally, contrary to biological theory and plain common sense, weasels and stoats actually tend to get smaller, not larger, the farther north they live—and thus even less energy-efficient. It sounds like a recipe for evolutionary suicide.

But according to King, weasels and stoats thrive or fail on the perilous edge of right-sizing: A weasel can afford to be small only because the runways down which it is thus able to travel usually lead straight to something juicy. Likewise, it needs no permanent home because it can invite itself so easily into Mouse House and Vole Hole, and then snuggle up after dinner in the grassy bed of its suddenly departed host. Being even smaller in the North makes sense, so the animal can spend the winter traveling down tunnels and burrows in the sheltered world under the snow. Milder winters in the South allow a slightly larger weasel to step out occasionally for a bit of rabbit, or other large game. Everything depends on being just the right size.

The weakness in the system is if there suddenly aren't any rodents at the end of the tunnel. When rodent populations go bust, which may be as often as every winter, weasel populations do, too. To adapt to this extremely unstable way of life, weasels have evolved unusual reproductive strategies.

The least weasel's strategy is more obvious than the stoat's. In the wild, least weasels typically live no more than a year, but a female can give birth and rear her litter to maturity in as little as three months. Unlike any other carnivore, she may then go on, in a good year, to produce a second litter. She can also vary the size of a litter, from zero to eight offspring, depending on what there is to eat. Her daughters, born in spring, can also produce litters of their own in summer. Thus, under ideal conditions, one female weasel in spring can yield thirty weasels by early fall. Then comes the inevitable rodent crash of winter, which can cause the weasel population to die back as fast as it has blossomed.

The strategy of stoats and long-tailed weasels, by contrast, sounds distinctly perverse. Stoats traditionally have been sex symbols of a sort in England. A lecherous man was sometimes described as "a bit of a stoat," perhaps by association with the suggestive shape or the frantic, lunging style of a stoat on the hunt. But science now suggests that the association is far more apt than anyone ever suspected.

Reproduction in stoats and long-tailed weasels is complicated by delayed implantation, a phenomenon best known in larger, longer-lived animals like seals and bears. The female gets pregnant in spring, the fertilized eggs then go into a state of suspended animation for perhaps ten months, and birth does not occur until the following spring. Since stoats and long-tailed weasels in the wild live a few years at best, delayed implantation sounds like an enormous waste of time.

King theorizes that the trait is a hand-me-down from the common ancestor of all weasels. Least weasels simply got rid of it as they evolved to take rapid advantage of rodent outbreaks. But stoats and long-tailed weasels stuck with delayed implantation. To compensate for the slower reproductive time, they developed what King calls the "extraordinary precocity of the juvenile females."

While the adult female is still nursing her litter, the male stoat comes to visit, often bearing gifts of food. Early observers, full of warm animal-story notions about the countryside, thought this was his way to help out with parenting. But this male is unlikely to be the father of the litter, conceived a year earlier, and he plays no part in child rearing. The gifts are intended, apparently, to elicit sexual favors, and this is where things start to get rocky for the anthropomorphic frame of mind.

Not only is the adult female ready to breed again while still nursing, but the daughters at her teats actually compete with her for the attentions of the visiting male. Nor is it merely the food they want. On his visit, the male mates with the mother and all her infant daughters. It isn't "an animal version of rape," said King. "They are positively keen, these little hussies." In one case, the party included a seventeen-day-old who was still deaf, blind, and toothless. But eager. Clearly (and if you have finished gasping, we can perhaps proceed), some lessons are best not learned from the natural world.

Likewise, it is best not to apply human values to other species. The combination of delayed implantation and female precocity makes sense for stoats and long-tailed weasels, according to King, because it means that almost all females are "pregnant"—that is, carrying fertilized eggs—almost all the time. In one study of five hundred female stoats, she found only two that were not impregnated. So if the rodent food supply goes bust one winter and there are no male stoats available the following spring, a lone female can still repopulate a habitat with her own offspring. In New Zealand, where stoats are considered a pest (having been introduced originally in an unsuccessful attempt to control rabbits), lone females sometimes swim out to island nature reserves, where their offspring are soon laying waste the native birds. If we can set aside human values for a moment, the stoat's strange sex life is simply a way for stoats to colonize and recolonize a boom-and-bust world.

I said good-bye to Carolyn King and Wytham Wood, and soon after I found myself on a road in the northern Midlands, driving past a menagerie of English pubs—The Hungry Horse, The Dandy Cock, The Beehive, The Swan. I was thinking about the strange twists of evolution, and the even stranger human notions about the world that evolution has given us. How odd that we should have heaped so much slander on the back of the weasel, or that farmers and gamekeepers should have expended so many centuries of murderous effort to get rid of them. Weasels are quirky characters, to be sure, but they are also a godsend to the countryside, as the bane of crop-raiding voles, rats, and other bucktoothed riffraff. They are the lionhearted heroes of our own backyards.

Just then, on a dark curve, I saw a sign outside a quiet pub. It depicted a fiddle-playing weasel in a red waistcoat, beating out the time with one foot. The name of the pub was The Waltzing Weasel, and it occurred to me that here perhaps the weasel could take its rightful place not as a creature of children's stories or wild mythology, but as a real animal. Here, if the patrons referred to a Hollywood producer as a "weasel," or a politician as a "stoat," they would mean it, in Carolyn King's words, as "something enormously satisfactory."

I decided to stop in at The Waltzing Weasel for a pint. But as I was passing through the door, a wave of retrograde thinking suddenly washed over me. I had a vision of the bartender with a compact mass of red parasitic worms wriggling behind his eyeballs, and of a Sweeney Todd sort of operation, in which, just about the time you get comfortable, the bar stool tips you down the chute into the meat grinder. I spun in my tracks and headed back out onto the road.

Old prejudices die hard. It was good to live in a world with weasels. Even so, I wasn't sure I wanted to have a drink with one.

Sharks, Part One:
Great White Hunter

In the stillness before dawn, while gear is still being stowed on board *Seaweed Too,* a thirty-six-foot charter fishing boat, Captain Greg Dubrule flips on the videocassette recorder to play his latest tape. It is, in effect, a series of commercials for himself—other customers on other days hauling in giant tuna, school tuna, pollack, bluefish, and of course shark.

It is also a sort of menu, and the question before this morning's customers, a party of six born-again Christians, is what to catch. Do they shoot for giant tuna, the glory fish, and risk getting nothing? Or should they pile up the blues and cod to earn back some of the $650 this outing will cost them? The consensus is to try for something bigger, something that smacks of sportfishing. Shark, in fact, sounds perfect. It also happens to be Dubrule's specialty: On a piling next to his berth, eight shark tails are nailed up like downed enemy aircraft, and one of them has a two-foot span. The prospect fills his customers with bravado and trepidation.

"We're gonna catch a lot of sharks today 'cause God's on our side," one young man declares. But when the videocassette shows the serrated mouth of a fresh-caught mako, someone else in the party, whose faith is not so certain, remarks, "Let's go for flatfish."

The *Seaweed Too* is under way at 5:20 A.M., rumbling out of the snug harbor of Noank, Connecticut. On a buoy outside the marina, a cormorant spreads its wings to dry in the new sun. Dubrule pushes the throttle forward, throwing up a deep wake like a whale's flukes, and heads out for the sharking grounds. The year is 1987.

Four years before, Dubrule and another Noank charter captain, Ernie Celotto, came back at the end of just such a summer day with a twenty-eight-hundred-pound great white shark in tow. It was fifteen and a half feet long and ten feet in girth, and it had been feeding on a whale carcass southeast of Block Island. The shark was front-page news ("Jaws monster landed after 10-hour battle," said the *Boston Herald*), and Dubrule told the story on *Good Morning, America*. The two captains donated the carcass to the University of Connecticut, which had a taxidermist make a mold for its new Museum of Natural History, and then Celotto returned to his charter fishing routine. But Dubrule had tasted celebrity, and he set out to build a reputation for himself as an East Coast shark hunter.

At age thirty-five, reed thin, in Topsiders and white corduroys, he is too young and well-scrubbed to resemble Quint, the seedy charter captain in *Jaws*. But more than a decade after Peter Benchley published his novel, the *Jaws* phenomenon obviously looms large for Dubrule and his customers. Indeed, it's probably the reason you can find shark-fishing and people eager to try it in Noank or just about any other harbor from Cape Cod south. So if Dubrule doesn't precisely fit the image, still, he knows enough to cater to it, giving his customers the requisite aura of danger within the reassuring context of a fully paid insurance policy. He wears a jawline beard in the manner of Ahab, and the back of his yellow T-shirt shows a wooden boat speared by a marlin like an olive on a toothpick.

There's also something Quintish about the rough way Dubrule handles his customers. When he remarks, with today's party assembled around him, that "these guys are green as grass," all six of them—three middle-aged men and three in their early twenties—shift uncomfortably, but no one demurs. They have not come here to be mollycoddled; this isn't just a fishing trip but a test: man against nature, man (with luck) against shark.

Dubrule makes it clear that he himself is not green as grass. He has been traveling these waters since childhood, when his family came down from a Hartford suburb to summer in Niantic. In high school and college, he worked on lobster boats and charters, learning the trade. He had his own fourteen-foot aluminum boat then and, while still a teenager, charged $50 a tide to take sportsmen night-fishing for striped bass, inshore. His father, an independent claims adjuster, tried to turn him away from the sea. (In fact, he is still pushing him to enter the family business.) But Dubrule says chartering is all he wanted to do.

He does it now from March through November, as many as three hundred trips a year, often working on only a few hours' sleep. Instead of an 18-horsepower outboard, he has a 300-horsepower diesel. Up on the bridge, he is surrounded with radar, fish finders, dual tracking auto Loran C, a seawater temperature gauge, and other high-tech tools of the trade. In his mind the rolling expanse of ocean is a map crowded with fishing spots.

"The more successful the fisherman, the more spots he knows," Dubrule says, as the *Seaweed Too* heads offshore at 14 knots, "and the more times of the tide to be on the structure. A professional has hundreds of spots—nooks, crannies, rocks, ledges. I fish on some spots that are no bigger than half the size of this cockpit. The Loran will get you to within thirty feet of the spot, and then you pick up the rest on the sonar—what customers call the fish-finding machine."

He considers this last phrase for a moment, and then adds, "You tell people it's broken, and they panic. They think you can't fish without a fish-finding machine. I try to tell them that *I'm* the fish-finding machine. Most of them have no comprehension of the chess game that's played up here."

Somewhere southeast of Montauk, Long Island, in 45 fathoms of water, Dubrule cuts the engine, comes down to the cockpit, and fills a net bag with chum—ground-up mackerel and bunker. "This is our calling card to the shark kingdom," he says, lowering the bag over the side. A slick begins to drift away from the boat, and petrels and shearwaters appear almost immediately to feed on it. Dubrule raises a steadying sail to push the boat along and spread the slick. His mate meanwhile sets out hooks, each baited with half a bunker, at depths of eighteen and forty feet. Dubrule looks

Shark Shopping for Dinner

out over the water, evidently satisfied, and remarks, "Now the wait begins." To his passengers, he adds, "If you fall in, kick. It'll help the slick."

Between June and October, the waters off southern New England are home to more than a half-dozen species of large shark. The blue shark, ranging up to 400 pounds, is the most common, followed by the sandbar or brown shark (up to 220 pounds), the mako (up to 1,000 pounds), the dusky (1,000 pounds), the tiger shark (1,000 pounds), and the hammerhead (300 pounds). Basking sharks, up to 5,000 pounds, are also common, but as the name suggests, they are a listless, plankton-eating species of no interest to sportsfishermen, except that they are sometimes mistaken for great whites. Sandbar and dusky sharks often come within ten miles of shore. The others mainly occur farther out on the continental shelf. But late in the season fishermen sometimes find the smaller makos and blue sharks in Block Island Sound and at the mouth of Narragansett Bay. Dubrule says he has caught sharks several hundred pounds in size within earshot of swimmers at Rhode Island's Misquamicut Beach.

While all of these species deserve respect, it is of course the great white shark—the so-called man-eater, the killer shark, the "*Jaws* monster"—that fascinates people, especially swimmers. Dubrule likes to say that, after Australia, the East Coast of the United States has the second largest white shark population in the world. This is open to debate; the area may simply have more shark fishermen. Even so, the average catch for the entire mid-Atlantic Bight, between Cape Hatteras and Cape Cod, is fewer than fifty great whites a year. Jack Casey, a shark expert with the National Marine Fisheries Service, says most weigh under two hundred pounds.

In his office in Narragansett, Rhode Island, Casey has a photo of the largest white ever taken, off Cuba in the 1940s. It was estimated to be twenty-one feet long and to weigh seven thousand pounds. But even big sharks don't normally eat people. Having dissected thousands of them, Casey has yet to find human remains in a shark. Fewer than a half-dozen shark attacks on humans have occurred in New England waters, and only one, in Buzzards Bay, Massachusetts, in 1936, was fatal. Both Dubrule and Casey say you have a better chance of being hit by lightning, or of being killed by a car on the way to the beach, than of being bitten by a shark once you get there.

The great white that was to become Dubrule's great white was first spotted by Ernie Celotto, who had a party offshore on his boat *Reelin'*. He tried to take the fish with rod and reel, but great whites are notoriously difficult to bait. Celotto sized up the fish, which was more than half the length of his wooden boat, and headed in at the end of the day intent on returning with a backup. *Seaweed Too,* he says now, "definitely wasn't my first choice." But Dubrule was eager.

"I'd been trying to get a big white shark for years," Dubrule says, and he makes no pretense about his motives. Ahab wanted the white whale to avenge the loss of his leg, and, in the movie *Jaws,* Quint killed sharks because they'd killed his fellow sailors on a ship sunk in World War II. But Dubrule says he wanted a great white "simply to draw attention to myself and my charter boat and then sit back and see what other avenues opened up. I knew that if it was properly handled, advertising avenues would open up. That was the sole intention of the fish."

At sea on the afternoon of August 5, 1983, Dubrule got his chance. The two boats were positioned on either side of the whale carcass, and the shark came up close enough for Dubrule's mate to plant a harpoon hilt-deep, with a sixty-inch float attached.

"There was four hundred feet of line between the shark and the ball," says Dubrule, "and the ball was underwater for twenty minutes. That's power."

Celotto says that when the shark surfaced, his boat landed a second harpoon behind the head. The two boats then fought the fish for the next five hours, catching up the line and holding on as long as they could, and putting on additional floats when they had to let go. Near the end, with Dubrule's boat hanging on, the line suddenly went slack. Down in the cockpit, the mate started to shout, "It's coming, it's coming! Get the hell out of here!" Dubrule hit the throttle, and as the boat pulled away, the shark broke the surface in its wake, rising several feet out of the water. "There's no doubt if we'd stayed there it would have come over the gunwales and landed right in the cockpit," Dubrule said afterward. It was like a scene out of Benchley's novel. They finally killed the shark with blasts from a 12-gauge shotgun.

For the first hour, only the petrels and shearwaters show any interest in the chum. The petrels have a way of skipping across the slick, as if walking on water. Every now and then Dubrule sweetens the mix, throwing in a chunk of bluefish and studying the water surface. He whistles and raps on the side of the boat, as if to call up the sharks. "Every other type of fishing, you know what you're going to get," he says. "Out here, you never know." To his mate, he says, "Come on, Mark, talk to 'em." Mark makes a kissing sound to the sea, then says, "Come on, babies."

Dubrule is conscious of the need not just to catch fish, but to keep his customers entertained. During long waits, he will sometimes turn on a stereo and play the theme from *Jaws*. Sometimes he will bring out shotguns and shoot skeet. "Anything new," he says, "we've got it." For $50, he will videotape the trip. He is also taking scuba lessons with the idea of videotaping sharks from a cage under the boat. He thinks of himself, he says, as "a multitalented enterprise, attacking from all angles." It is a competitive business. "You've got to stay one step ahead."

At 12:15, the party gets its first shark on the line, and the customer fighting the fish with rod and reel is soon down on his knees. "Next time he comes up, somebody shoot the sonuvabitch," he gasps. But in twenty minutes, an eight-foot-long blue shark is struggling at the stern. Mark, the mate, reaches down and grabs the line too forcefully.

"Don't hold that line," shouts Dubrule, who is poised to stick the fish with a harpoon. "The muscle man here. You're going to pop something!" He jabs the fish once and then again, but the dart doesn't penetrate. On a third try, Mark grabs the line again and the fish rolls just as the dart drives home. The wooden shaft snaps against the boat, an ungraceful finish, and Dubrule is furious. "Steady pressure," he yells. "Whatever pressure the man with the rod is putting on, that's what you put on when you grab the line. If you jerk it, he'll take off again, and who's fault is it? The mate's. You've got to have finesseful hands at the boat."

The night Dubrule and Celotto brought their great white back to the dock, seventy people were waiting, and others continued to show up all the next day. Among them was Peter Benchley, who posed for a photograph next to the carcass. "There is no sport to harpooning a fish," he declared. "It's like going out and shooting dogs."

Nowadays, Dubrule looks back on Benchley's surprise attack philosophically, one promoter appreciating another. "*Jaws 3* was coming out then," he says, "and it was the perfect avenue for him."

But at the time, the two fishermen were bitter. If sharks were being killed wantonly, as Benchley said, the blame rested largely with Benchley himself. Celotto told reporters that, after *Jaws* came out, "we had businessmen with attaché cases stepping on board." Inside would be a .45 automatic. "All they wanted to do was go out and shoot a shark." A biologist in the National Marine

Fisheries Service shark-tagging program defended Celotto from the charge of wanton killing, and both captains said that they typically tagged and released far more sharks than they killed.

At a press conference, Dubrule attempted to take the high road: "There is no money to be made off the shark," he said. "We didn't have a personal gain in mind. We're just two hardworking fishermen. We are donating it to science for teaching purposes."

And in fact the shark was to prove unusually valuable to science. Research on the animal's dental bacteria subsequently led to the first pharmaceutical therapy for sharkbite victims. Dubrule continues to regard this as an important defense: "Anybody that gives us any grief about needlessly killing the animal, if they can't buy the sportfishing end, they'll have to buy the medicinal end."

The great white's carcass was handed over to Carl Rettenmeyer, a University of Connecticut biologist, who needed a centerpiece for the Museum of Natural History then being planned. Rettenmeyer put up $2,000 of his own money to fly in taxidermists and make a fiberglass mold. ("It was a Sunday and we had a rotting fish on our hands," he says.) Rettenmeyer also gave the two captains a letter valuing their contribution for tax purposes at $10,000.

Dubrule meanwhile began exploring the promotional avenues opened up by his great white. "Before," he says, "I always imagined I'd spend my life on the boat taking out parties. It involves a lot of hours, more than a hundred hours a week, and a lot of cooperation and understanding from a wife and two children at home. You miss out on a lot of things in life." But the shark promised to change all that.

Dubrule found himself in demand for off-season fishing shows. "You can go to someplace like Wisconsin where they've never seen one of these things before," he says, "and, my God, they make a hero out of you—'famous East Coast shark hunter.' This isn't *Jaws* that they've got to relate to through Benchley. This is the real McCoy. I can make more money on that than I can clear in a day on this boat. I get $300 to $500 a day. But the thing is, we're a main attraction."

At 1:27, Dubrule's party hooks up its second shark. "Bless God," the angler cries. When he has been struggling for some time, the mate remarks, "The fish doesn't even know he's been hooked yet." Dubrule is more encouraging. "When you can gain, gain," he advises. "Just keep it tight. Let him do the work. When you put out an ounce of energy, get a pound back."

The angler goes down on his knees (for reasons that have nothing to do with religion), and another member of the party shouts: "Don't lose it now, you woman." (A curious epithet, as Dubrule has just finished talking about a woman angler who held onto a four-hundred-pound shark for four and a half hours, reduced at times to tears but never to the point of surrender.) This shark, another blue, comes up to the side of the boat at fifteen minutes, and, with the angler's consent, Dubrule tags it behind the dorsal fin and sets it free. (The tag, in a small capsule, gives the date and place of the catch. When the shark is taken again, off Cape Cod or perhaps Florida, the information will help scientists chart migration patterns.)

Almost immediately, the third and final shark of the day takes a bait and starts to run.

When Dubrule says "we're a main attraction," he means himself and the shark. At the booth he sets up at fishing shows, he has a full-size fiberglass mount of the great white on display. Sometimes he brings a fighting chair, too, and sits one member of the audience there with rod and reel while another grabs the end of the line and plays shark. It is a good show. But it doesn't make either Carl Rettenmeyer or Ernie Celotto terribly happy.

According to Rettenmeyer, Dubrule probably shouldn't have the mount in the first place. Having donated the carcass, he says, Dubrule then went to Rettenmeyer's taxidermist and arranged to buy the first mount made from the fiberglass mold. Rettenmeyer delayed ordering his own mount because the Museum of Natural History wasn't ready yet, nor did it have the $2,500 additional cost. "We essentially paid for the mold for his shark," says Rettenmeyer.

Dubrule replies that the arrangement with Rettenmeyer was "just a slimy deal. I had an oral contract with that sonuvabitch that we were going to donate the mount to the museum on the condition that we could use it in the winter. I'm a promoter in my own right. There's no way I'd give away something that I've been trying to get for so long without getting something in return."

The dispute between them comes down now to quibbling over who has the better mount. Dubrule says his mount came first, and "in the eyes of the public, that is the taxidermist's mount of my great white." Rettenmeyer, whose mount went on display when the Museum of Natural History opened in 1986, says his has the real teeth and more accurate detailing.

Celotto's unhappiness has less to do with the mount than with Dubrule's entire presentation. "Once the carnival atmosphere started, I backed away," he says. "I didn't want to build my reputation on the capture of one shark. I don't choose to be the Barnum & Bailey sort." Perhaps more to the point, he objects to the way Dubrule refers to the shark as "my shark," narrating its capture without reference to the second harpoon or the second boat.

Dubrule replies: "Ernie Celotto had nothing to do with it from the start. He saw it, that's all. He watched us catch it. Ernie Celotto was there in spirit only."

As his customer winds in the third shark of the day, Dubrule suggests that they tag and release this one as well. "I want it," the angler says, through gritted teeth.

This is of course the customer's privilege, but Dubrule points out that blue sharks aren't such good eating and that they don't freeze well. "How much can you eat fresh?" he asks. "You've got a lot of meat here already. I don't think you fully realize . . ."

It develops that one member of the party, a church pastor, keeps a pair of pit bulls. "They love fish," says the pastor.

"If you're just going to feed the dogs," Dubrule says, "I'll give you some bunker."

But it turns out that the shark, another blue, is already mortally wounded. It comes to the side of the boat after just six minutes, with its stomach hanging out of its mouth; it has swallowed the hook. Dubrule finds two other hooks in the shark's mouth, one of them recent enough to have a large piece of mackerel hanging on to it.

"They say that a fish wouldn't get caught if it learned to keep its mouth shut," the pastor remarks. "You'd think this one would've learned."

But there is plainly something troubling about the catch, or perhaps about their own appetite for killing sharks. In the cabin afterward, the party talks over the day's haul uneasily: Who's going to clean them? Who's going to lug the sharks up and load them onto the car? "The mate'll carve 'em up and give us fillets," someone replies. Then, a little defensively, he adds, "There's fifty pounds of meat on that fish. How much does fifty pounds of dog food cost?"

Still later, a member of the party holds up a severed shark tail and someone says, "You can hang that from your rear-view mirror." Then they set to work cutting out the jaws, which are the real trophies.

On a snowy night in Worcester, Massachusetts, six months later, Dubrule and the mount of his great white occupy a corner booth at the Eastern Fishing and Outdoor Exposition. A videotape Dubrule made in August shows a great white shark chomping methodically on a whale carcass. "Note the nice, clean mark it leaves in the side of the whale," Dubrule tells his audience. "The neat, circular shape. These are the largest and most feared predators on the face of the earth today."

People take away copies of the *Seaweed Too* brochure, but during a lull, Dubrule confides: "A lot of people think I'm here drumming up charter business. In fact, I've got all the charter business I can handle. I'm here for exposure. I've got a real feature attraction here that it's easy to build an advertising campaign around. I don't care if it's pantyhose, Coca-Cola, or sportfishing rods. Look at the potential of this thing."

He brings out a flyer produced by a Connecticut auto dealer for an upcoming promotion. It shows Dubrule's shark with its jagged mouth gaping, over the headline "Mallon's RV takes a big bite out of RV prices for '87." Dubrule and his mount will be there, in person. "Hear Captain Greg Dubrule describe his incredible life or death fight," the flyer urges.

A young man comes up to Dubrule's booth and, indicating the shark mount, inquires casually, "That the one that was biting on boats a few years ago?"

"That's the one," says Dubrule. "Had a sixteen-foot Starcraft in its stomach at the autopsy."

"Oh yeah?" says the young man.

Dubrule is indeed good at this sort of thing. He is, as he says, a salesman, a showman. This year, he has improved his act with a genuine shark jaw taken from another twenty-eight-hundred-pound great white, which his boat harpooned by itself in August. When a crowd gathers, he lifts it overhead like the holy grail and says, "I'm holding the jaw of a three-thousand-pound white shark." He picks a young woman out of the crowd. "Let's see if this fits around you," he wonders. And as the thing engulfs her torso, he says, "Yup. One bite, and good night, Irene."

Dubrule and his shark have even the ushers at the show talking. "It's almost like a small whale," says one. And another replies, "It's almost like a tearing motion." He pulls at the air with both hands. "What a way to go!"

Unfortunately for Dubrule, there is a second, subtler shark presence at this show, though not in person. A tackle manufacturer's booth gives the details: This past August, Captain Frank Mundus and his angler, Donnie Braddock, out of Montauk, caught a 3,427-pound great white shark, not with a harpoon but with ordinary tackle, using a line with a tested breaking strength of 128 pounds. It was the largest fish ever taken on rod and reel, and they landed it in an hour and forty-five minutes.

Mundus has been catching sharks since the year Dubrule was born; he was the model for Benchley's Captain Quint, and many people consider him the real McCoy. This isn't to suggest that he's averse to commercialism; it turns out that he has sold one of several mounts of his shark to Ripley's Believe It or Not, which will also display a wax figure of Mundus himself. But over the years, Mundus has become averse to what he calls "sticking" fish, taking them with harpoons.

Reached in Hawaii, where he is sitting out the fish stories of midwinter, he says the real sport with great whites is to bait them. "You have to tease 'em and tease 'em and tease 'em," he says. He does it from a platform that projects fourteen feet off the side of his boat, where he stands and swings the bait like a pendulum. When the white shark passes below, he throws the bait just ahead of it, and then pulls it away. The teasing can go on for hours. "Once you get movement from his eyes you know you have him. As he gets close, you pull it away, till he gets so mad. . . . The timing has to be just right, because if you give it to him before he really wants it, he won't take it." Harpooning, by contrast, is simple. The great white moves too slowly, sometimes poking its head out of the water right next to the boat; it is too easy a target. "A stuck fish, it's nothing," Mundus says.

At the show in Worcester, it's a slow night. Other charter boat captains drift by to chat with Dubrule about business. It is, as one of them says, a line of work where "a lot of money passes through your fingers, but not much of it sticks." They talk about last season's luck, both good and bad, and about the prospects for the new year. The tape of the shark eating the whale carcass plays over and over in the background. "Mundus might've got the shark," Dubrule remarks at one point, "but we got the video." To another captain, he says that with their knowledge of the sea, and their many endeavors to extract a living from it, the two of them have something special, something marketable.

"If we could only direct it some way to make money," says Dubrule, "we'd be millionaires."

Sharks, Part Two:
How Sharks Got into
Such Deep Soup

On a glorious October morning not long ago, a surfer named Rick Gruzinsky was drifting off the Hawaiian island of Oahu, gently rising and falling as he waited for his wave. A large green sea turtle paddled by. There was a light breeze. Gruzinsky noticed colors and swirls underneath him. He thought it might be the turtle again, or a shallow coral head. Then the shark hit, lifting the surfboard out of the water and flipping it.

With its teeth sunk into the bow rail, the shark, a fourteen-foot tiger, shook its head once, and then again, snapping off a chunk of fiberglass. "I remember distinctly seeing the eye just below the water level and the big round snout," Gruzinsky told a reporter later. "The shark was trying to swallow the piece, and I remember looking into . . . the soft white part of the mouth." Then the shark sank slowly out of sight, leaving Gruzinsky, who suffered only scratches, to paddle 150 yards to the beach.

Two weeks later, a shark picked an eighteen-year-old out of a group of body-boarders off Oahu. He bled to death on the beach.

Earlier in the year, also off Oahu, a surfer disappeared, leaving behind a board with a chunk bitten off, and a few months before that, a fifteen-foot tiger shark slammed into a snorkeler off Maui, sweeping away the victim's legs and one arm.

Public reaction to this series of attacks was in part predictable. Parents kept their children out of the water, and the legislature debated "tiger shark eradication." A travel industry official worried that, if it happened to a tourist, it would make every newspaper in the United States and Japan. Advocates for a shark kill noted that Hawaii's last control program ended in 1976, time for a new generation of sharks to grow big enough to eat people. The state Department of Land and Natural Resources sent out a posse to kill large sharks in areas where attacks occurred.

But these are strange times, and the outcry on behalf of the sharks was at least as loud. Some scientists argued that a control program would be futile: The rate of attack was determined not by how many sharks you took out of the ocean, they said, but by how many people you put into it. They also pointed out that while 20 million people a year use the beaches in Hawaii, the rate of attack remains minuscule—2.4 per year. Better to spend the money, they said, for more lifeguards, to reduce the annual toll of thirty or forty deaths by drowning.

The public also took up the shark's defense, even more passionately. "We think of them as family," said one protester from a Hawaiian family, for whom sharks were traditionally sacred. "They were there before us. That's their domain; it's their territory. Why pick on them?" Telephone threats caused one fisherman to give up the shark hunt. Under pressure, the state ultimately diverted most of its shark-hunt money to research, conceding that it didn't know how many sharks were out there, where they came from, how long they stayed around Hawaii, or what made them attack.

Even Gruzinsky was ambivalent, when hunters dragged in the shark that had apparently attacked him: "It was a sad thing to see, an animal like that dead, as big as it was. . . . There has to be some way we can both enjoy the ocean."

If what happened in Hawaii was strange, it was also typical of mixed feelings now surfacing everywhere people live with sharks. Because of overfishing in the 1980s, shark populations have apparently plummeted around the world. People who used to sit on the beach and fret that there might be sharks out there now fret that there might *not*. New and improbable-sounding ideas aimed at saving sharks and finding out how they live abound. But in the debate over whether to kill sharks or to protect them, it is literally the nature of the beast that neither side knows for sure what it's talking about.

South Africa, for example, recently declared the great white shark a protected species and imposed a ban on fishing for them or selling their jaws or other parts. In the coastal province of Natal, which has been thinning out its shark population for forty years with a system of fishing nets off major swimming beaches, officials announced that they would release the 10 to 20 percent of sharks that survive capture in the nets, including great whites.

These measures were doubly remarkable, in a country with a rich history of shark attacks on humans, because proponents readily admitted they had no evidence of a decline in great white sharks. The aim, according to Leonard Compagno, a shark biologist in Cape Town, was to prevent local waters from turning into a shark "free-fire zone." He cited the case of a single fisherman off the Western Cape who took eighteen large white sharks, extracted their jaws for sale on the American market at as much as $5,000 each, and ditched the carcasses. Australia has also recently banned trade in great white parts, and restricted fishing for several other shark species. Seven shark species are now on the IUCN–World Conservation League's Red List of threatened species.

In the United States, federal officials in 1993 imposed a long-delayed fisheries management plan along the Atlantic and Gulf coasts, aimed at the formerly unthinkable goal of *rebuilding* the population of sharks. The plan came from the National Marine Fisheries Service (NMFS, pronounced "nymphs"), which traditionally regarded sharks as a plague. It aimed to limit the catch of thirty-nine shark species. For recreational fishermen, it imposed a bag limit of four sharks per boat per trip. Among other restrictions, the management plan also banned the notorious practice of stripping off the fins, which are the most valuable part of the shark, and dumping the carcass. In 1997, NMFS made the management plan even tougher, reducing the commercial haul of large coastal sharks by about half over two years.

How did such a fierce and unlikely creature as the shark become the object of so much concern? In the aftermath of the *Jaws* movies, shark populations around the world came under attack. Recreational "monster fishing" tournaments proliferated, with the first prize sometimes worth as much as $60,000, and extra prizes for the boat weighing in with the highest overall shark poundage.

Commercial fishing boats also began targeting shark, as tuna and swordfish stocks ran down. Shark meat had been the stuff of scandal when Louisiana schools were caught feeding it to children in the 1970s. But through the magic of marketing, it suddenly became a hot item at restaurants.

At about the same time, liberalization in the People's Republic of China led to the end of long-standing strictures against the consumption of shark-fin soup, which can sell for as much as $150 a bowl. The immediate surge in demand—and price—for fins drove shark fisheries around the world to a frenzy. Because the price of shark meat remained relatively low—or worthless, for some species—fishermen often salvaged only the fins and threw back the carcass. Some people thus liken what happened to sharks in the late 1980s to the slaughter of the buffalo in the 1870s. Fishermen dispute this analogy, but hardly anyone denies that sharks have gotten harder to find.

"In the early 1980s," a sport fisherman named Allen Ogle told me one night, as we headed seaward off the west coast of Florida, "you usually had to have a fish in the seven hundred pounds to have a chance of winning a shark tournament." That number has dropped so quickly that in 1990, the top fish in one Florida tournament was a ninety-two-pound nurse shark, a species sometimes derided as "overgrown catfish."

Out in sixty feet of water, Ogle baited a couple of big, rusty hooks with a sawed-off hunk of tarpon the size of a Christmas roast. The boat was alone in the dark between untroubled sky and sea, with the faint amber glow of Sarasota low down in the east. He drew out ten feet of line, picked up the leader, and slung his offering out over the gunwale, where it disappeared with a cannonball splash into the empty sea.

"I'm targeting big ones with this," he said. "Eight feet and up." The thrill of big sharks stripping line off the reel is what got him into tournament fishing in the early 1980s.

But the thrill quickly faded. "Guys got discouraged," Ogle said. Because urea and other compounds in the blood make shark meat inedible unless it is immediately butchered and refrigerated, almost all tournament fish wound up rotting in landfills or being towed back out to sink—a level of waste that made some participants uneasy, and fearful for the future of their sport. "There was a change in attitude after the first couple of years hauling in big fish and hanging 'em up," Ogle said.

Odd as it may sound, tonight's event is thus a no-kill catch-and-release tournament for sharks. Mote Marine Laboratory, a Sarasota institution specializing in shark research, sponsors the event and boasts that "only the information is landed." Contestants are obliged to bring the shark to the side of the boat long enough to measure it and plant a research tag near the dorsal fin. For every foot of shark, they earn a ticket for the first prize lottery. Then they set the shark free.

It is a noble idea whose time may already have passed. Ogle said he typically waits thirty hours to get a run now, where an hour or two would once have sufficed. At three in the morning, he hauled in the unmolested hunk of tarpon and headed home, having detected not the faintest *dunt*-dun-*dunt*-dun tremor of a shark on the line.

Environmental groups, notably the Center for Marine Conservation in Washington, D.C., have pushed hardest for protecting sharks. The U.S. NMFS plan has also won endorsement from the Sportfishing Institute, the shark research community, coastal state governments, and the industry-oriented regional fisheries management councils.

But when the draft plan first appeared in 1989, the shark industry reacted with outrage, in part for reasons familiar from other disputed fisheries. Sharks migrate freely across international borders, but the management plan could do nothing, for example, to limit the Mexican shark industry, which is double the size of the American fishery and fourth largest in the world. "All we're doing is penalizing Americans, and not doing anything to restrict the import of shark fins, shark teeth, and shark skins, and to me, that's a travesty," said Sonya Girard, a Louisiana fin trader, who is nonetheless savvy enough about her business to have set up a Latin American import operation.

The fishermen also professed outrage because it was NMFS that got them into the business in the mid-1980s, with seminars and videotapes touting sharks as an "underutilized" resource. "They made it very easy for me to start this business," Girard said. "They gave us the names of Chinese people who dealt in fins; they gave us their fax numbers." Hence NMFS didn't have much credibility with fishermen when it turned around two or three years later and told them that sharks weren't, after all, underutilized, but perilously overfished.

"They get their figures out of the sky," said Chris Brannon, whose Alabama seafood company made a reputation for taking more sharks faster than anybody else. NMFS originally based its proposed quotas on estimates that there were 814,500 large coastal sharks left on the Atlantic and Gulf shores in 1987, and

3,737,000 small coastal sharks in 1989. "How in the hell do they know all that?" Brannon inquired. He argued that researchers were leaping too eagerly onto the "Save the Shark" bandwagon.

The final NMFS plan revised these estimates upward and doubled the fishing quota, angering environmentalists. "I have a problem with the numbers, because they run counter to all the assessments that have been done," said Sonya Fordham, a fisheries specialist at the Center for Marine Conservation. "I imagine it all comes down to politics." But the final plan, she added, was "better than what we have now, which is nothing."

The truth, of course, is that nobody knows for sure how many sharks are out there. NMFS researchers come up with their numbers by sampling an area, or by examining catch records, and extrapolating from these data with the help of mathematical models. Because almost everyone mistrusts such models, both sides wind up pitching "our" war stories against "theirs."

In 1989, for example, Grant Gilmore, a researcher at Harbor Branch Oceanographic Institution in Florida, videotaped sandtiger sharks at a wreck off the North Carolina coast. On his video, the massive blue shapes fill the screen, six or ten animals at a time in the immediate range of the lens, moving with a languorous flicking of the tail. As the camera pans, the sharks appear everywhere, endlessly. But when Gilmore returned to the site in 1991 for research, he found it empty. Local divers told him that an out-of-state fleet fishing for sharks had taken everything. (In fact, they blamed it on Brannon, who denied responsibility.)

Every shark researcher on the East Coast tells a similar story, and they argue that even if their population numbers are merely best estimates, it makes sense now to err on the side of caution. "If you waited till you had all the information you need, you might not have a fishery," said a NMFS administrator. "They'd be extinct."

Except for one or two species, this information is woefully lacking: Researchers don't know for sure how fast different sharks grow, where they live, how often they reproduce or in what quantity—in short, all the data needed to figure out how to harvest a resource without destroying it.

"People get this image of this one shark, the *Jaws* image," said Sonny Gruber, a longtime shark researcher at the University of Miami. "But if you look at the three hundred seventy or so species of shark, some are as different from one another as we are from rabbits." All sharks share such defining traits as straplike gills and a cartilaginous frame instead of bone. But beyond that, said Gruber, "There are sharks that are flat as pancakes, sharks that have beards and whiskers, sharks that glow in the dark. There are sharks that swim up to whales and bite plugs out of them. They're called cookie-cutter sharks, and they have these suctorial mouths; they spin like a top and take a plug out of them the size of a silver dollar." There are sharks that feed cooperatively, slapping mackerel or bluefish out of the water with their elongated tails, and there are sharks that patrol the bottom in solitude, using bioelectric sensors unequaled in the animal kingdom to detect the telltale heartbeat of flatfish hidden under the sand. Indeed, even as they discuss the possibility of extinction for some sharks, scientists are still discovering three to five unknown shark species each year. "In the study of sharks," said one researcher, "we are where people who study birds were in 1900."

One night, in a broad, mangrove-lined lagoon on the island of Bimini, fifty miles off Miami, I joined a group of volunteers wading waist-deep through a shark nursery. Big, pregnant lemon sharks had already come and gone in the shallow waters here, each leaving a litter of about 10 fully formed young. Their offspring will stay here for the first year or two of their lives, preying on small fish and crustaceans around the roots of the mangroves, where bigger fish are less likely to prey on them.

The volunteers work for Sonny Gruber, and they quickly learn to speak of themselves as having been "Gruberized": Their leader is a wiry, demanding figure in a bush hat and sleeveless khaki shirt. His manner is weary and relentlessly busy, mixing laid-back "Ya, mon" island talk with abrasive commentary on the petty details of the operation, down to how thick the lunch bread is being sliced and whether a visiting BBC television crew is taking sodas without paying for them (an inquiry proves them not guilty). The clock in the dining room is prominently labeled: "For organizational reference only. We live as the shark," which seems for Gruber to mean that constant movement is a biological requirement. But if shark work sometimes breeds eccentricity, the possibility of working with sharks also makes volunteers tolerant of it.

Out in the lagoon, an associate named Charlie Manire has positioned long nets across the tidal flow, and it is the volunteers' job to lurch back and forth across the spongy bottom for twelve hours seeing what they catch. I watched one of them reach in to grab the pale-bellied, comma shape of a two-foot lemon shark tangled in the net. In front of us, a glittering, eight-inch-long fish rolled up to the surface, with a red stump where the shark had chopped its tail off. The volunteer placed one hand firmly behind the shark's dorsal fin and guided it through the water back to a waiting boat. With its broad, rounded gray head, it looked like a pet nuclear submarine.

The idea was to catch every newborn lemon shark in this lagoon, over three nights, and then repeat the exercise in a year to see how many survive under relatively natural circumstances. Preliminary results indicated that 80 percent may die before their first birthday.

Like most large sharks, lemons grow slowly, taking twelve to fifteen years to reach maturity. Like us, they reproduce by copulation and internal fertilization, allowing the female to provide greater nourishment and protection, but for a small number of embryos. An individual cod or flounder can lay millions of eggs per year, to be fertilized externally. The lemon shark needs a full year to produce a single litter, and then waits till the following year before she mates again. (Another shark, the spiny dogfish, has the longest gestation known for any animal, almost two years.) In lemon sharks and other species that nourish their fetuses with a placenta, the umbilical cord actually leaves a small scar between the pectoral fins, the equivalent of a belly button.

Before anyone leaps to sentimentalize sharks, though, they should consider the *in utero* world of the sandtiger shark. In the sandtiger, the first embryo breaks out of its elastic, egglike capsule when it is four inches long and swims around the uterus, ripping open other developing egg capsules to devour the embryo inside. According to Grant Gilmore, who discovered the phenomenon, the sandtiger has two uteri, and the strongest offspring in each works its way systematically through its siblings. At birth, the two survivors are forty inches long and amply prepared by the comforts of the womb for the rigors of the outside world.

By whatever means—infant mortality in lemon sharks, intrauterine cannibalism in sandtigers—shark reproductive strategy tends to produce relatively few adults. This is a good thing normally, because adult sharks are at the top of the food chain, and have few natural enemies. But when humans use drift nets, fish-finders, six-hundred-hook longlines, and other modern technology to take out huge numbers of adults, sharks, unlike cod, lack the ability to replace themselves quickly.

"Nobody can make a living killing sharks," said Gruber. "You can't make money with slow-growing pelagic fish that you hammer down so they don't have a breeding stock." He held up his fingers like a closed flower and then burst them apart. "*Poof!* It's finished. Fifty years to recover."

Like other children of the 1930s and 1940s, Gruber remembers being force-fed cod-liver oil for vitamin A. The medicine actually came from sharks, in which the disproportionate liver is a kind of buoyancy chamber brimming with oil. (Shark-liver oil, properly labeled, is still a major ingredient in Preparation H.) In California, demand for vitamin A drove the price of soupfin shark livers from $11 a ton in 1938 to $1,653 just four years later. Synthetic vitamin A replaced shark-liver oil in 1950, but the West Coast population of soupfin sharks has yet to recover. The literature is full of similar cases. Peddlers of New Age medicines, for example, have contributed to a recent boom in demand for sharks; despite the total absence of scientific proof, they tout shark cartilage pills as a treatment for cancer. As a conservation-minded shark fisherman told me, "There's never been a shark fishery that hasn't crashed."

So why did NMFS promote a shark fishery on the East Coast? "Remember where we were," said Paul Leach, chairman of the task force preparing the shark management plan. In the early 1980s, NMFS was clamping down on overfishing of king mackerel and snapper-grouper. But sharks appeared to be everywhere. Fishermen were not only catching them unintentionally on their longlines, but also found that sharks were tearing into the tuna and swordfish they hooked up. "That's what was taking a good forty percent of your catch," said Chris Brannon, the Alabama seafood entrepreneur. "NMFS' attitude was, 'Hell, catch 'em. Let's develop a fishery.'"

Targeting sharks also had undeniable emotional appeal. According to the most reliable estimate, sharks attack fifty to seventy-five people a year worldwide, causing five to ten deaths—an incidence rare enough that one researcher has a sign on his wall advising: "AVOID SHARK ATTACKS—Get hit by lightning." But the ingrained human dread of sharks persists. Ernest Hemingway sometimes strafed them with a Thompson submachine gun from his boat. J. L. B. Smith, one of the great marine scientists of this century, once wrote: "We must seek to evolve devices that will attract and ruthlessly destroy every shark that comes anywhere near."

Gruber regarded this as no more than paranoid wishful thinking. "I never believed what J. L. B. Smith proposed was possible." Then people began catching sharks for crab bait in the Florida Keys, just as Gruber was beginning a study there in 1986. "The first year we got about a hundred and fifty sharks, the second year about a hundred and ten with the same effort, the third year, thirty-five sharks. In 1989, we got only seven sharks, and then we realized the study was over. That's when I wrote my first shark conservation paper. Before that, I didn't think the problem was real."

NMFS also did not think the prospect of a shark population crash was real. It was public opinion, rather than research, which ultimately changed its mind. "It took finning," said one NMFS researcher, "before the public outcry and the emotional level reached the point that we decided the fishery was in peril."

In one widely publicized case, two Florida fishermen were caught in 1988 killing bottlenose dolphins and using their meat as a shark attractant. Federal agents prosecuted them for violating the Marine Mammals Protection Act. But they also noted that when the pair caught sharks, they sliced the fins off the living animals and left them to die, and that this was entirely legal. It was the first time most Americans ever heard of a market for shark fins.

Shark fins are almost devoid of meat and, to your average Burger King customer, utterly unappetizing. They contain yellowish collagen fiber in between the cartilage, which looks like a bird's nest of noodles when first extracted. In ancient China, noble families gauged one another's social status in part according to how well the family cook prepared this fiber into a gelatinous soup. Chefs who botched the job sometimes lost their heads.

In modern China, shark fins are still a symbol of power, and an ingredient in folk remedies. According to Oregon fisheries consultant Sid Cook, a handful of families operating out of Hong Kong control the trade worldwide, in an atmosphere of high secrecy. They turned to the United States in the mid-1980s after having already depleted shark fisheries in the Arabian Sea, West

Africa, and Central America. Prices here promptly shot up from $4 for a pound of "wet," or unprocessed, fins in 1985, to as much as $14, and the gold rush began.

The shark industry is still trying to live down the bad reputation it got from the fin trade and from a handful of incidents in which sportfishermen actually caught sharks after they'd been finned. Live-finning, said Sonya Girard, the Louisiana fin trader, "was the emotional catalyst that started the management plan, and I'm trying to explain to you that it just didn't happen. Sharks are awesome animals, and you don't just say "Scuse me, I wanna take your fins.' "

But Girard added that many sharks, such as hammerheads, arrive at the surface dead, after having been immobilized on a longline hook. "Hammerhead meat is too mealy and too pink; consumers don't want it," she said. "If you don't have a market for it, what do you do with it?" It would be wasteful *not* to take the fins, she suggested. Under the fisheries management plan, American boats are only allowed to bring the fins into port with the rest of the body still attached—a rule which should make finning economically impractical.

But no one has yet developed hooks or nets that discriminate against hammerheads and other undesirable "bycatch." Too many sharks will thus continue to be caught by boats that don't want them or aren't equipped to deal with them. NMFS estimates that in 1988, U.S. fishermen were discarding 90 percent of the sharks they caught, taking only the fins and reserving refrigerator space for higher-priced tuna and swordfish. Other nations are probably no better. An Australian researcher recently reported that, along the Tasmanian coast alone, Japanese longline vessels targeting bluefin tuna inadvertently catch 43,500 blue sharks a year. "The sharks are discarded after removal of the fins," the researcher noted, no market for blue shark meat having yet been devised. Sharks also die in huge numbers as a bycatch of high-seas drift-netting. A multinational team of observers on Japanese squid boats recently recorded that drift nets took 58,100 blue sharks in a single six-month period. Unlike hammerheads, blue sharks often come to the surface alive, leading Jack Casey, a NMFS researcher, to suggest that live-finning may be more common than Girard would like to believe: "It's dangerous. It's tricky," he said, "but if the price is right, fishermen are pretty resourceful in terms of restraining an animal." Until he retired, Casey ran a long-term shark-tagging program, and a commercial fishing boat recently recovered a tag he put on a blue shark twenty-five years ago. Then the crew finned the shark and heaved the live body overboard.

Among the many things nobody knows about sharks, one of the most disturbing is what an ocean without sharks would be like. Some prey, such as the stingrays favored by hammerheads, would probably boom. "Godalmighty, there are stingrays!" Allen Ogle, the reformed tournament fisherman, complained to me at one point, as we drifted in his boat out on the Gulf of Mexico. "We had sixteen different swimmers hospitalized for stingrays off one beach early this summer in Sarasota. When you think about that," he added, with the earnestness of the convert, "it's easy to see that the shark can be our friend, too." In Australia, ecologists speculate that the rise in the shark fishery may have led to the collapse of the spiny lobster industry in some areas. With the sharks gone, a variety of small octopus apparently flourished, and preyed on the lobsters.

Sharks may also influence other species in more subtle ways. Like wolves, they tend to pick off weak or injured animals, leaving the quick and the strong to reproduce. In French Frigate Shoals in northwestern Hawaii, for example, fledgling albatrosses routinely land on water after their first flight. Tiger sharks patrol the nearby lagoon in anticipation. When a shark makes its attack, head and open mouth breach the surface, and it happens often enough that one researcher observed 138 attacks in a single summer. (One of the problems with the idea of tiger shark eradication, researchers theorize, is that fledging season may attract sharks from thousands of miles away. The resurgence of monk seals and green turtles may also cause migrating sharks to linger around the islands.) The tiger sharks manage to kill 10 percent of albatrosses leaving the rookery. Rocky Strong, a Cousteau Society researcher, theorizes that such attacks select out animals on the basis of intelligence, agility, and other survival traits. The cruel logic of predation is that it improves the prey.

One afternoon, I went out shark-tagging in the Florida Keys with Bill Botten, a genial, balding retiree, who wore tortoiseshell glasses and a "shark sushi" T-shirt depicting humans wrapped in seaweed. Poling his skiff across the flats, Botten scanned the shallow water and said, "There's one cutting across that little mangrove island there!" He cast once, said "Ah! Wrong side," and again, more accurately. In a moment, a young lemon shark was in the boat, tagged and measured for the NMFS shark survey, and set free again. Botten has tagged more than eight hundred young sharks in the Keys over the last decade, though fewer and fewer in recent years.

But as we fished, he suggested that it was naive to concentrate only on the plight of the shark, that it smacked of the American tendency to skip from issue to issue in the manner of the news magazines: timber wolves this week, harp seals the next. He worried about the entire marine environment, and shook his head with familiar late twentieth-century trepidation. "I think it's pretty broad-based," he said, "I just hope they're not the canaries in the coal mine."

I cast in front of a lemon shark thirty feet away and watched it catch the scent, and make its tight, erratic advance. When I cast a second time and pulled the bait past, the shark sprang on it. I reeled it in, and the dorsal coming toward me across the water was ominous even at that scale. It occurred to me that the nice thing about sharks—in a decidedly abstract, third-person sort of way—is that they improve *our* species, too. They keep us literally on our toes, looking around us for shadows in the surf. They serve as toothy reminders that in the natural scheme of things we were not put on this earth, or in the oceans, to rule unquestioned. In movies, literature, the paintings of Winslow Homer and John Singleton Copley—and at least until now, in real life—sharks have always served as a small check on our overweening pride. Botten planted a tag behind the dorsal fin, and I lowered the shark into the water again. It lay in my hands for a moment, and then surged free, back to the dwindling safety of its natural world.

A Porcupine Would
Rather Be Left Alone

Wendell Dodge, a lean, taciturn New Englander in a checked timber jacket, plants one foot forward on a rock and prepares to croon a porcupine love song. He leans back, his hands in the pockets of his blue jeans, looking like an aging Lothario at a promising balcony. The object of his attentions is an ancient, mossy white oak with a hollowed-out trunk just ten feet away.

Dodge, a wildlife biologist now retired from the U.S. Fish and Wildlife Service, has had luck here before. There is an opening near the base of the tree, with a sort of stoop outside the door formed by generations of porcupine droppings. But the porcupine population in this part of western Massachusetts has fallen off drastically, he says, because the trees have not been "throwing mast" (producing acorns and other nuts) at their usual rate. One way to tell if any porcupines still make their home in this particular den is to serenade it.

Dodge murmurs the first note almost inaudibly, but his voice rises steadily in both volume and fervor—*hmmmm, HMmmmm,*

173

Porcupine

HMMMMMmmm—then trails off in a quavering, hopeful, slightly disappointed diminuendo. He gets no response at first. He repeats the call, a sound made from the top of the throat, with mouth closed and lips tight. It is like a baby whimpering from a dream that is bad and getting worse, and on the third try it touches the heart of something inside the white oak. A low, scrabbling sound emerges, as of claws on wood. Dodge calls again. Then, lower and much more tentatively, something replies: *Hmmmm, hmmmm, hmmmm.*

The duet has begun, and whatever is in the white oak keeps up its end for several minutes without appearing. Tension builds. Dodge moves closer, and he and the white oak continue mewling back and forth with each other. He reaches down and rustles together a handful of leaves, the porcupine equivalent of bedsheets. The voice within is caught between terror and curiosity, perhaps even desire. Then another scratching sound, and suddenly a black ball trundles headfirst down the ramp and out onto the stoop. It rises on its hind legs, sniffs the air, and declares itself: *Hmmmm, hmmmm, hmmmm.* Behold *Erethizon dorsatum,* literally "the beast with a formidable back."

• • •

The porcupine is a very likable animal, for a rodent. It rarely bites, though its incisors can clip through a hemlock branch as cleanly as pruning shears. It almost never attacks, though it can defend itself formidably. There is nothing cunning or predatory about it. Porcupines sustain their ten- to twelve-pound bulk quite nicely on a vegetarian diet of tree parts (the cambium and phloem beneath the bark), nuts, other plant products, and the occasional sheet of plywood. After food, their fondest wish is to be left alone under a ledge or in the branches of a tree. The porcupine is, in fact, timid.

This trait may at first seem utterly at odds with its chief physical characteristic, the 15,000 to 30,000 quills that prickle forth from each animal. This spiny protective coat begins at the brow, where the quills bristle forward like a cowlick in a butch haircut. The quills become thicker and as much as three or four inches long across the hump of the back, and then shorter out to the tip of the tail. Only the belly and the face are unprotected. When threatened, the porcupine turns its back and seems simultaneously to cower and bristle, its head down and its quills clattering nervously together like spears.

The quills are loosely attached, and any animal foolish enough to touch them will come away with a faceful. Quills have been found on polar bears (the porcupine's North American range includes much of Alaska), deer, owls, a twelve-hundred-pound Black Angus cow, dogs, even a trout (the porcupine will sometimes swim to harvest a choice water lily). The porcupine cannot hurl its quills, despite the capabilities of cartoon porcupines. But by thrashing its clublike tail against an antagonist, it can drive its quills a good half-inch into the flesh, Dodge says.

To the unaided eye, the quill is merely a wickedly sharp toothpick, with a white shaft and a black tip. But a microscope shows that the pointy end is shingled with hundreds of tiny barbules, while the center of the shaft is a spongy material. As the quill penetrates the skin, the spongy material soaks up liquids and expands, causing the barbules to flange out. Every time the victim's muscles contract, they catch on the barbules and draw the quill deeper. The quill has a penetration rate of about an inch a day, a statistic determined by a selfless naturalist who left a quill stuck in one side of his calf and counted the days till it worked its way through to the other side. Dodge, who says he is not that devoted to science, once got sixty-three quills in the leg; he removed them all. The pain of insertion isn't that bad, he says, "hardly worse than jamming a sliver in your hand." But pulling them out is agony.

The quills allow the porcupine to lead a subdued, solitary life, indifferent to most other animals and untroubled by its own serious physical weaknesses. Porcupines cannot see the danger in front of their faces. Dodge has known porcupines so myopic or so lovelorn that, in response to his mating call, they not only emerged from their dens but actually reared up inquiringly with their forepaws on his leg. Porcupines have sensitive hearing, especially to low-level sounds like a footfall. Their sense of smell is also acute; the porcupine's snout is all nostrils. But even if a porcupine hears or smells danger, it isn't much good at running away. On the ground it lumbers along, high-stepping awkwardly with its forepaws. It is quick only on the pivot, whirling around to keep danger at its back. In the trees it climbs like a bear, with its semi-prehensile forepaws on the sides of the trunk, and moves slowly enough for any predator to catch up. Only its permanent rear guard allows it to get safely to the top.

Porcupines are not gregarious animals. To put it in terms of strict scientific precision, they are a stiff-necked, unsociable species: Their cervical vertebrae are actually fused together. Porcupines will sometimes share a den with other porcupines in the winter, but not happily. Dodge says his white oak has frequently been the scene of "squawking and strife." When they quarrel (and perhaps also when, and if, they cuddle), porcupines must pause intermittently to pluck out foreign quills with their forepaws and teeth. All things considered, they would rather be left alone. The single gaudy exception to this solitary existence occurs during the mating season, which may last from September till as late as January. This is the only time Dodge has found two adult porcupines up the same tree. Folklore has the porcupines making love face-to-face while suspended like trapeze artists by their forepaws. In fact, porcupines find arboreal romance awkward and almost always adjourn to the ground. Their precoital rituals are unusual. According to Dodge, the male spends a considerable amount of time making an olfactory examination of the trysting place and marking it with his own scent. Then the male rises on his hind legs and, from a distance of as much as fifteen feet, anoints the female with bursts of urine. If the female is not receptive, says Dodge, "she'll screech and caterwaul like a bobcat." Otherwise, after further play, mating proceeds conventionally. The female flattens her quills and arches her tail over her back. The male mounts from the rear, resting his forearms on the underside of her tail or letting them hang down loosely at her sides. After a pause for grooming, the couple may tussle again briefly. The interlude ends, according to Dodge, in "hostile screaming and lunging." The wonder is that it ever began at all.

If all this suggests that porcupines are an undesirable animal, that's okay with Wendell Dodge and just about everyone else who studies porcupines. Their publications on the animal inevitably make prominent use of the word "control." As far as humans are concerned, the trouble with porcupines is mainly a matter of plywood in their diets. Not to mention canoe paddles, steering wheels, rubber tires, brake linings, automobile and tractor hydraulic systems, sugar maple tubing, tree plantations, and highway signposts. The porcupine's chisel-like incisors never stop growing, and like other rodents, it uses them assiduously to gnaw things. Its jaw muscles make up about half the weight of its head.

Despite what this bulk and its diverse menu might suggest about the creature's mental capabilities, the porcupine does not exercise these muscles indiscriminately. Porcupines have taste, of a sort. Given three kinds of apples, a porcupine will almost always prefer the red Delicious (to the chagrin of parsimonious apple farmers, who also favor the variety as a cash crop). If apples aren't in fruit, it will still prefer the red Delicious tree for its bark. Porcupines also like plum trees. And they will climb cornstalks and flatten them to get at the ripening kernels. So why do they bother with unappetizing stuff like tires and canoe paddles? The reason, apparently, is that they also crave salt, whether from salt spreaders on the highways or from human sweat. The result, agriculturally and otherwise, is that the porcupine can make itself a costly nuisance.

This sort of thing (to say nothing of the exotic sexual behavior) does not go down well in New England, and porcupines there have never enjoyed the popular status of, say, the armadillo in Texas. They were particularly unpopular earlier in this century, when they returned to reforested areas ahead of their natural predators and consequently boomed. John Barrows, a district forester with the state of Vermont, recalls that Vermont used to offer a bounty of fifty cents for a set of porcupine ears, and in 1952 paid out $90,000. Remarkably, it still had a porcupine problem in 1953 and for several decades thereafter. Barrows explains: "There was a time when we thought the state had a lot of money, and a trapper who knew how to use his knife could get ten or twelve sets of ears out of a single animal."

Vermont thus did not really get hold of its porcupine problem until it began to airlift in a predator who worked for love rather than money—the fisher, a member of the weasel family once native to the state. The fisher is to the porcupine what the mongoose is to the cobra. It goes in low and quick, slashing tooth and claw at the unprotected face. The fisher has lately flourished to such an extent that it has a trapping season of its own, for its fine sable coat. A natural balance has ensued, and even lumber companies can now afford to tolerate the diminished porcupine population. No one should fret, however, for the porcupine's survival. It fends for itself; Barrows has seen a fisher blinded and choked with quills, dragging itself off to die.

There are two things most people would not think of doing with a porcupine. The first is to have it for a pet, and the second is to have it for dinner. But both practices are surprisingly common, considering the hazards.

When Dodge was a graduate student, one of his animals could not raise her newborn. His wife Polly adopted the pup (known technically as a "porcupette") and began feeding it formula at home. They named it Jim, and it sometimes went to school with Dodge and followed him around the campus. It would sit on a desk or, when each was secure enough about the other's good wishes, on Dodge's lap. Jim liked lollipops, which he would hold in his forepaws and eat while sitting up on his hind legs. He also enjoyed ice cream cones, and Dodge says the animal would eat them on the way home from the parlor while sitting on the rear ledge of the family's '53 Ford. "It used to startle motorists," he reflects. Then one day Jim suddenly realized he was a porcupine, turned ugly, and had to be given to a zoo.

Barrows has also seen porcupines as pets and says they are at least as trainable as dogs. Porcupines have the advantage, he adds, of not being stupid enough to poke their faces around where they are likely to get a face full of quills. This is a common avocation of New England dogs, who are also stupid enough, on having two hundred quills painfully extracted, to go out and seek revenge.

A further bit of lore has it that at one time several western states gave the porcupine protected status—not, however, because it was in any sense endangered. The idea was that the porcupines were so nutritious and so easy to kill that they were worth preserving as a survival resource for people stranded in the wilderness. But even in death, even on a spit or between slices of white bread, porcu-pines can fend for themselves. In one pertinent case, a man was admitted to a hospital with severe stomach pains. Surgery showed him to be suffering from a porcupine quill through the intestinal wall. It turned out that he had eaten a carelessly prepared porcu-pine sandwich four days before. The damage done by the steadily penetrating quill killed him. In its own eerie, characteristically pas-sive way, the porcupine had gotten its revenge. What a strange notion, that such an animal as the porcupine would need *any-body's* protection.

Notes from the Underground

It was a cold, damp March morning in the Border Country of northern England. Snow lingered along the field walls, making white stripes across the green Northumberland hills. Lapwings and curlews tumbled overhead. Peter Rutherford, professional mole catcher, was advancing across enemy territory, a pasture pocky with hundreds of mole diggings. He surveyed the wreckage for activity, and then whispered, "There may be one puttin' up a heap over there."

He crept forward in his muddy boots. With a loose stocking cap on his head and the shovel strapped across his back, he looked like a sapper in the army of King James reconnoitering some ancient fortification. He gestured with his hand to hold up. After a tense moment poised with shovel ready, Rutherford suddenly plunged forward to the attack.

The blade sliced into the earth on one side of the mole heap, and Rutherford sprang round and jammed his heel down to close off the tunnel on the other. In another instant, he had the heap and a six-foot line of suspected escape routes dug out, muttering, "Come on, son, where the hell are ye?" He dropped to his knees and ran his hand along the excavated banks, feeling for side tunnels.

"There he is!" he yelled, indicating some imperceptible quaking in the soil. And then, ferociously, "YER KNACKERED!" His hand pursued the unseen mole into the earth. "Where the bloody hell are ye now?" he demanded. "Ye bastard." He was up to his elbow in the dirt. His face went red with vexation and doubt.

Then, triumphantly, Rutherford pulled his hand from under the bank. In his grip was a squirming mole, a female six inches long, her pink paddle-feet scrabbling against the unfamiliar air. Her mouth gaped in outrage, exposing a long narrow jawline of needle-wicked teeth. She shrieked like a crow, a sound alarmingly at odds with the meek and genteel image of Mole, the celebrated character in Kenneth Grahame's children's classic, *The Wind in the Willows*. "She's a bit upset," Rutherford said, as if apologizing for an old acquaintance. Then he turned on her venomously, "Ye *little* sod, ye!"

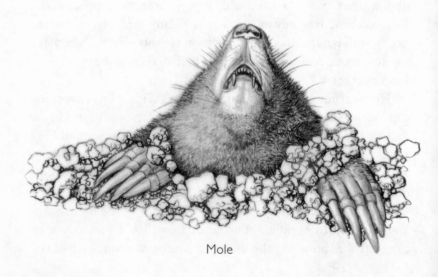

Mole

A little spoiler of the sod, a subterranean wrecker of lawns and pastures—that's the mole's reputation, and the source of Rutherford's livelihood. With its spongy tunnels and its ability to ravage a fairway or front yard with willy-nilly two-foot-high heaps of dirt, the mole has a knack, as a British writer recently put it, for driving mild country parsons to prime their shotguns, and bank managers to grin malevolently over their traps. The mole has burrowed not just into our yards, but under our skin, and become an object of murderous outrage.

People set traps designed to kill a mole by exploding a .32 cartridge in its face, by spearing it, or by strangling it in a scissors grip. They put broken glass, razor blades, thorns, mothballs, and exhaust fumes down mole tunnels. They invest in sonic repelling devices that use precisely the high-frequency sound waves that die out most rapidly in dirt, evoking indifference, if not scorn, among the moles. About as effective, but less humane, is the recent British innovation of placing musical Christmas cards within the tunnel system, presumably to drive the mole insane by endless replayings of "Little Drummer Boy."

Outside of children's literature, you have to look a long way to find a kind word about moles—and even when you find one, it has to do with destruction: Certain Scots still fondly recall that King William of Orange, cruel usurper of their own King James II, died from a fall when his horse stumbled on a molehill. They sometimes drink a toast to "the little gentleman in the black velvet waistcoat."

Otherwise, human attitudes toward moles constitute a lengthy chronicle of contempt. Moles have customarily been regarded as disgruntled loners, the "raging psychopaths of the countryside," according to mammalogist Martyn Gorman, inclined to fling themselves furiously at any living thing unlucky enough to cross their trail, including other moles. The mole has plenty of reason to be angry, common wisdom tells us, because fate has trapped it in an endless, insatiable round of digging and eating. Its only blessing is that it is blind. That way, it can't see what a lousy life it leads, "ever destined to live in darkness," as Oliver Goldsmith put it, entombed from birth to death in its grave.

Out of the ground, the mole might well *look* contemptible. Depending on the species, its eyes are either completely hidden or barely visible and perceive little more than changes in light level. Its pink, uplifted nose and its meandering side-to-side movements bespeak shortsighted uncertainty. The way the head widens without any visible neck into the massive upper body musculature suggests not just a strong back, but its corollary, a weak mind. Children's books usually depict the mole as a kind of absent-minded professor. In truth, it looks more like a no-neck, third-string nose tackle who's lost his contact lenses at the line of scrimmage but is nonetheless hell-bent on doing damage. On appearances alone, it is entirely understandable that a wretched landowner would see ample cause to belittle the creature that is actually driving him insane. One morning in Northumberland, Rutherford pointed out 143 dead moles hung up by a farmer in a grim row along a strand of barbed wire, as if to say, "Take that, you bloody little rodent!"

But moles have a way of confounding our assumptions. For one thing, they aren't rodents at all, but insectivores, like shrews and hedgehogs. The moles are a small group; there are only about thirty species in the world. North America has seven species, from the Townsend's mole (*Scapanus townsendii*), which bedevils dairy farmers in Oregon's Tillamook cheese country, to the star-nosed mole (*Condylura cristata*) in the Northeast, which looks as if it's got a sea anemone stuck on its snout.

The moles come from a venerable lineage; the fossil record of their recognizable ancestors dates back about forty-five million years—far earlier than our own direct forebears—and moles appear to have made good evolutionary use of their head start. Researchers have begun to demonstrate that, far from being the literal wretched of the earth, moles may, in fact, be perfectly adapted—astonishing miracle!—to live precisely the way they live. Martyn Gorman, of the University of Aberdeen in Scotland, argues that moles actually lead "rather charmed lives" underground. He studies the common European mole (*Talpa europaea*), which has become the archetype for moles in children's books, scientific research, and the historical fur trade.

Most of the mole's obvious adaptations serve to make it a better digger. Other burrowing mammals dig with their forepaws underneath them, in doggy-paddle fashion. But the mole's pectoral girdle benefits from a dramatic rethinking: The humerus, equivalent to our upper arm bone, angles up and out from the body; it's also flat and wide, the better to attach large digging muscles. This setup causes the mole's forelimbs to stick out to the sides like gates, and it maximizes the power of the sweeping stroke the mole uses to scrape soil from the sides of its tunnel and push it rearward, or up to the surface. Some moles also have an extra bone in each paw, for a broader shoveling surface.

This equipment makes moles prodigious diggers and is the source of their cartoon reputation for churning across lawns before some outraged human's eyes. A mole can dig sixty feet or more of subsurface tunnel in a day—roughly equivalent to a five-foot-tall woman burrowing the length of two football fields. (To compete with the mole, by the way, the woman would also have to be able to move objects in her path, up to, say, four thousand pounds. One thinks of Carla, the waitress in the television show *Cheers*.)

The mole has evolved the means to manage its work load while breathing rank, carbon dioxide–laden tunnel air, the equivalent of what humans exhale. A fifth of its body weight is given over to its voluminous lungs. Its blood is also adapted to its way of life, almost visibly. When a mole like the one Peter Rutherford plucked from the ground meets its usual demise—by a sharp rap on the head—gouts of blood as bright and saturated as enamel drip from its earholes. Moles contain twice as much blood as other mammals their size, packed with twice the amount of oxygen-bearing hemoglobin.

So far this might not seem to rival Paris in the 1920s in the category of charmed lives. Indeed, in their book *The Natural History of Moles,* Gorman and coauthor R. David Stone give a subchapter the plaintive title "Can Moles Do Anything Else But Dig?" The answer is, yes, they can sleep, and they spend half the day at it. To confirm this, Gorman and his associates radio-tracked moles for the first time, using Superglue to attach micro-transmitters to their subjects' rumps.

For two weeks, one of Gorman's students, Sara Frears, tiptoed around a field in Northumberland once every hour, recording where each radio-tagged mole was located underfoot and what it seemed to be doing. In between, Frears holed up from the January west wind in her unheated van, thinking enviously about the underground way of life Mole describes so fondly in *The Wind in the Willows:* "Nothing can happen to you, and nothing can get at you. You're entirely your own master. . . . Things go on all the same overhead and you let 'em." Frears notes that the temperature is relatively constant underground, and that her snug little moles did not need to grow a thicker fur coat, lay on fat, alter their metabolic rate, or make any of the other customary physiologic concessions to winter. (Frears herself subsequently made the ultimate concession, migrating to New Zealand.)

Frears was interested in how moles use energy, and her calculations suggest that even digging isn't as hard for a mole as it's cracked up to be. A mole burns only about five times as much energy digging as when it's resting. By contrast, a bat's metabolic rate spikes up by ten to thirty times when it flies. Moreover, while flying is the basic means of gathering food for a bat, digging isn't for a mole. Moles spend only about two and a half hours a day digging in the peak tunneling season of winter, and only a few minutes a day in summer.

And yet the mole is a voracious eater. It consumes about half its weight daily, and it can take fifteen to twenty juicy night crawlers to knock the edge off a day's appetite. (Frears once kept ten moles in the lab, and found that digging up enough worms to feed them was a full-time job. An American researcher complains that his three moles ran through $17 worth of worms in just two days.) So where does its food normally come from?

The genius of the tunnel system is that, once established, it brings the prey to the predator. Worms literally drop in, as an English countrywoman puts it, "like bacon and eggs on the ceiling." Once they land in the tunnel, worms and insects tend to stay there. It may be that the surfaces are so compacted and polished by the mole that it discourages escape. In any case, all the mole has to do for a meal is amble periodically around its estate gathering up the latest crop of house guests.

This larder is so reliable that the mole carries around only about 2 percent of body weight in fat, enough to keep itself alive for less than a day. "There's a cost to carrying around body fat," says Frears, especially when the animal has to squeeze its bulk through a tunnel of its own making. "And why pay the cost if food is readily and reliably available?"

Each mole patrols its own estate in solitude. These estates may overlap in places, but the moles live up to their reputation as nasty little loners. They've evolved a time-sharing strategy to avoid the horrid prospect of actually meeting up with a neighbor.

Moles make their rounds three times a day, with each activity period lasting three or four hours. They are acutely aware of their neighbors' movements, and all the moles in an area leave the nest and return to it at about the same time, like workers in a company town. But the moles appear to use their overlapping tunnels only once a day, discreetly avoiding the area when another mole is present.

Early accounts of the mole's ferocity relied largely on reports that two moles placed in the same box often fought to the death. But in their own world, moles are more subtle. They communicate their presence and avoid brutal confrontations by scent-marking on their daily rounds. Because the worms in their diet are largely water, moles urinate abundantly, which may make it easier for them to mark several hundred yards of tunnel a day.

"Chasing season" is the one exception to their fierce isolationism, when a switch suddenly flips somewhere in the male and he goes burrowing off in a beeline until he breaks into the tunnel system of "an attractive and welcoming female." The phrase comes from Gorman and Stone, who should perhaps know better. The female is generally hostile. Indeed, Gorman and Stone speculate elsewhere that her ovaries may actually produce male hormones for much of the year. This might explain not just her aggressiveness, but her external resemblance to the male. Except during the mating season, the vagina is sealed shut, and the urinary projection closely resembles the penis of the male, who also lacks a scrotum. Mating appears to be brief, violent, and possibly confused. Peter Rutherford reports that you can hear the caterwauling rising up from underground, and it doesn't sound like splendor in the grass.

The female wastes little time on mating or family life, but the latter is at least quality time: Her milk boosts the three or four youngsters in her annual litter to seventeen times their birthweight in three weeks. Then they get the heave-ho, and everyone goes off to the chief business of moley life, which is gathering worms in blissful subterranean solitude.

When a mole finds a worm, it generally pins the body between its forepaws while pulling up on the head with its teeth. Squeezing the worm in this fashion, like a tube of toothpaste, may help clean the surface and push out soil the worm has eaten. The mole gobbles its meal with an eager, almost ecstatic quivering of the whole body. But it doesn't necessarily bolt down everything it meets.

Sometimes the mole merely nips off a worm's first two or three segments. Then it balls the worm up and rolls it back to the nest, to be stored for future use. Out in the Northumberland hills one morning, Peter Rutherford discovered a nest mound, a pile of dirt slightly larger than the waste heaps moles had put up all over the field. He opened it with his shovel and quickly turned out the mole's bed of dry grass. A network of tunnels ran through the mound, and Rutherford speculated that some of them were designed for drainage, to divert water away from the nest.

Then, along the gallery tunnels of the mound, Rutherford started to unearth stored worms. Each one was coiled in a loose knot and tucked into its own honeycomb pocket, immobilized but still alive. "When you take one out, it'll dig itself into the ground again," he said. "So you wonder, why don't they just nick off? See how they're pushed into little round pockets? I think the mole must seal them in with clay balls to stop them." He plucked up the coiled worms and tossed them in a pile on the grass. For a mole catcher, as for a mole researcher, gathering worms is half the job. Rutherford would stir this lot of worms with poison and redistribute them in mole tunnel systems all around the farm. "I've struck it rich," he said, still digging. "There's bloody hundreds."

The coiled worms were like the living dead, slightly diminished in bulk and paler than usual—some of them almost opalescent. "Even though they have their noses cut off, they keep on shiting," said Rutherford. "And they shite all the muck out of them."

He gathered up the cold, wriggling mess of worms in his cupped hands, 133 in all. Other investigators have found up to 1,280 worms, weighing more than four pounds, cached in a single nest.

Having mixed the worms with poison, Rutherford headed into a new pasture, remarking, "Look at the bloody mouldies in here!" He surveyed the blackened heaps, deciphering which territories were active and which tunnels were major routes. Then he plunged a wooden daub stick into the ground, finding the hollow of a tunnel as deftly as a phlebotomist seeking a blood vein. With a twig, he lifted a worm from his bucket and dropped it into the tunnel. Then, because moles are sensitive to any intrusion, he closed the hole with his heel. "The next time he comes along, he'll get a bellyache," he said.

Its bumbling image to the contrary, the mole is extraordinarily alert, in its own world, to the faintest hint of trouble. The poison was strychnine, which can be perilous even when applied underground. Dying moles are sometimes found and eaten by other predators. But strychnine is the only poison now known that fools the mole's keen sense of smell. A mole will also "push a trap right out of the ground," said Rutherford, if it detects human smell.

The sense of vibration may be even more acute. The mole cruises its tunnels with its tail touching the ceiling like the overhead pickup on a trolley car. The tail, head, and feet are lined with sensory vibrissae—rigid hairs set in fluid-filled sacs to amplify the slightest mechanical stimulus. Even when it's digging so furiously that you can hear the sound of roots tearing, the mole seems to sense approaching danger.

The Rutherford family once had a spaniel that could creep up and point on a mole working underground. But when a plodding human came close, the mole would fly to the cover of a deep tunnel. Rutherford recalls the dog looking back witheringly at its clumsy master, as if to say, "Ye idiot." Few predators get close enough to kill a mole, one of the central charms of underground life. Those that do may wish they hadn't. Gorman has an X ray of a herring gull that ate a mole. The mole proceeded to dig a hole in the gull's stomach, shoulder its way through the wish bone, and get its head halfway out the gull's neck before both animals expired.

Fixated as we are on the sense of vision, humans can scarcely imagine the world as the mole perceives it. The mole's snout is covered with a cobbled, fleshy structure that contains highly sensitive receptors called eimer's organs, which flush red in the presence of food. A researcher enthuses that the eimer's organs are "unexcelled in complexity in the whole animal kingdom," and then adds that no one knows for sure what they do. In one experiment, a molelike animal called the desman could distinguish a full food container from an empty one by using its eimer's organs to "read" lines engraved, like incredibly minute braille, on the lids. Another researcher speculates that the mole may use its eimer's organs to detect thermal emissions from worms, which are slightly warmer than the soil around them.

Few animals make better use of their snouts than the star-nosed mole. This bizarre North American species split off from its fellow moles about thirty million years ago, according to Terry Yates, who studies mole evolution at the University of New Mexico. Star-nosed moles can dig like other moles, but to exploit their marshy habitat, they've also evolved in a way that enables them to pursue their prey underwater. Yates writes that they swim even under the ice on frozen lakes and streams. The star-nosed mole's tail is long and ratlike, unlike any other mole's, and in the water it apparently functions as both rudder and paddle. Also unlike other species, the star-nosed mole lays on fat seasonally, perhaps because it ventures out into cold weather so often. The tail can double or even triple in thickness as the seasons change; it's an ingenious place for a burrowing animal to carry extra fat while still remaining streamlined for tunnel travel.

But it's the nose, not the tail, that gives this species its name. No snout in the mammal kingdom, not even the elephant's subtle trunk, is as mobile and complex. Around the star-nosed mole's nostrils, twenty-two tentacle-like appendages fan out, wriggling continuously as if to taste the air, water, or soil the animal is traveling through. The nubbly surface of these appendages contains altogether about twenty-five thousand eimer's organs. The nose is only about a half-inch across, but it is wired with more than a hundred thousand nerve fibers—roughly six times as much as the entire human hand. Researchers have generally theorized that the appendages of the nose serve primarily as touch receptors.

But Edwin Gould at the National Zoo in Washington, D.C., has suggested that the nose may also be a remarkable electrosensory organ for finding prey in muddy water. Worms emit an electrical field, and when Gould simulated this field with a buried penlight battery, the moles seemed to home in on the signal. He also observed that when a star-nosed mole swims, it swings its head from side to side. On picking up an electrical signal from a worm in the water, it lunges across the final inch or two. The bite typically lands on one of the electrical hot spots on the victim's body. This may be a serious blow for the image of the mole as a bumbling, faintly pathetic animal.

But the image, unlike the worm, will probably survive. Moles being largely unseen in real life, most people will continue to get to know them mainly from *The Wind in the Willows*. Since its publication in 1908, this tale of Mole and Toad rambling across the English countryside has endeared moles to millions of readers. It hasn't spared the moles from persecution. But nowadays they typically get affection along with annihilation. This is true even at the vast country estate near Henley-on-Thames that Kenneth Grahame used as his model for Toad Hall, the main setting for his story. "Luckily, they've most of them just been trapped," the lady of the house told a recent visitor, and then she felt obliged to add, "They really *are* adorable looking things. I wouldn't want them to become extinct. It's just that there are *so many*."

Indeed, there are so many that, improbable as it may sound to delicate modern sensibilities, moles once supported an entire industry. From the late nineteenth century until about World War II, moleskins and furs were an article of high fashion, in part because of the mole's special adaptations to underground life. Its hair moves readily in any direction, enabling the mole to advance or retreat with equal ease through a narrow passage. Furriers once used the soft four-inch-square skins to make moneybags and muffs, to trim coats, and to stitch together into whole garments.

King Edward VII started a rage for moleskin vests in 1903. It took seventy-five skins to make such a vest, according to one account, and it was customary for it to be joined together in front not with buttons but with a purple ribbon. In the United States, a 1939 Saks Fifth Avenue catalogue featured an elegant woman in a stylized lumber jacket, cinched in at the waist and peaked at the shoulders, made of several hundred moleskins.

At the height of the fashion, Europeans harvested twelve million moles in one year. Germany had to pass legislation to protect moles. But the mole survived, testimony to the success of its isolationist strategy. By reproducing at a low rate, taking advantage of a steady food source, and staying underground where even trappers couldn't get them all, moles populated and then repopulated Europe as thoroughly as humans.

The trade in mole furs is now almost extinct, as is the classic old-time mole catcher, who often bore an uncanny resemblance to the mole itself. Peter Rutherford recalled one such ancient Northumberland trapper, short, squat, and dark, who never went to bed, but sat up all night in front of a smoking fire, inadvertently acquiring a permanent mole-black patina. He also generally kept a heap of fifty or so moles on the kitchen table, not counting the stretched pelts on the wall. Visitors reported that he tended to skin the carcasses and butter his bread with the same knife.

This is a practice on which Peter Rutherford's girlfriend would frown. In any case, at age thirty-nine, with a sturdy build, small, deep-set eyes, and a full, flat honest face, Rutherford was too young to fit the image of the classic mole man. A former miner, he'd just switched over full-time to his father's old trade and was only eking out a living. "It's bloody hell getting money out of farmers," he said, wandering the hills one morning. His $80 bill for a day's work often goes unpaid long enough for Rutherford to contemplate enforcement measures: "I'm a good hand at catching live moles. So I can give them back."

A more serious problem for Rutherford was that some hill farmers suffered, if not from Wind in the Willowish sympathies, at least from a failure to see the practical benefit in controlling their moles. They were stuck on the ancient question of just how much damage moles actually do. Doubts on this subject crop up even among mole-murdering integrated pest-management specialists, the IPM guys, who have largely replaced old-fashioned mole catchers.

"People are very sympathetic to the mole. I mean, he probably does a lot of good, aerating the soil," said one, who works the Henley-on-Thames landscape where Mole did most of his wandering. "It's just that he's such an untidy animal, throwing the junk out on top. Poor old mole. It's discrimination." Vanity about our lawns, he added, is the real reason most people kill moles. Contrary to the deeply held belief of many gardeners, moles don't eat precious flower bulbs. Gardeners might do better to complain that moles eat earthworms that would otherwise be enriching their soil. But moles also eat some agricultural pests.

It appears that the only people with a legitimate gripe against moles are small farmers. Though moles don't actually eat crops, their tunnels can sometimes dry out an entire row of plant roots, or bring in mice, voles, or other pests. Their heaps can blacken a pasture, according to an Oregon study, which estimated that dairy farmers in Tillamook County could lose up to a third of their land to moles. British officials in the early 1980s blamed moles for about $3.3 million in annual agricultural damage, a relatively small sum, except perhaps to a farmer fretting away the last weeks of winter watching mole heaps erupt all across his land.

Rutherford pointed to a farm across the valley. "That's what I like to see," he said. "That bloke there, I did his fields about five years ago. I gave him the bill and he said, 'That's too much.' Now his fields are absolutely black, and serve the bastard right." He consoled himself with the thought that the moles would spread to the neighboring fields of sensible farmers, who would want them killed. There would never be a shortage of work. The persistence of moles, snug in their earthy homes, guaranteed it.

Then Rutherford told an Irish joke. His listener, being of Irish descent, chose to take it as a remark not on the Irish intellect, but on the charmed life of the mole.

"This rich Englishman woke up one morning," Rutherford said, "and there was a mole heap in his yard. He couldn't sleep for thinking about it, and he couldn't eat his dinner. So he advertized for a mole catcher.

"It was an Irishman who answered the ad, and he said, 'How much will you give me if I catch the mole?' And the Englishman said, 'I'll give you five hundred pounds, and when you catch him I want you to give him the worst death you can imagine, because of all the sleepless nights I've had worrying about him.'

"So the next day the Irishman showed up. 'I've caught your mole and I've come to collect my money.'

" 'And what death did you give him?' the Englishman asked.

" 'I gave him the worst death I could possibly imagine,' the Irishman replied. *'I buried him alive.'* "

Jungle Days

They were like scholars in a great museum that was crashing down around their ears. They worked among the disappearing treasures with efficiency, dispatch, and even humor, as if a lifelong student of antiquities could carefully record the exquisite details of a Greek amphora and then watch with equanimity as someone hammered it into dust; or as if the world's leading authority on Chippendale furniture could note the particulars, and then stand by as side chairs and scroll-top highboys were smashed to splinters around him.

What these researchers were cataloguing was only ostensibly less precious—one more rain forest being chainsawed into oblivion, with no more than the usual assortment of plants and animals produced by millions of years of evolution and found nowhere else on earth. The researchers were accustomed to this kind of destruction. They had spent much of their careers in the field studying forests like this one and seeing them vanish.

. . .

Up the trail, a botanist named Al Gentry follows a tape measure through the forest, calling out the names of species in a fifty-meter-long transect. Yesterday, a bulldozer cutting a logging road knocked down one end of the transect while Gentry was still working at the other. Now the bulldozer is once again pushing perilously close, and trees are crashing down nearby. Gentry, whose strategy toward all hazards is to pretend they don't exist, is hastening toward the finish when he spots something up a tree that he must collect. An assistant reaches up with a long, wobbly cutting pole to retrieve it, unsuccessfully. Since the tree is going to come down anyway for the logging road, the assistant begins to hack away with his machete. At the crucial moment, with the bulldozer operator yelling for them to move along and the tree beginning to tilt, Gentry loses his footing and slides down the wet slope.

"*Cuidado! Cuidado!*" someone shouts. "Watch out!"

The tree comes thumping down beside him, close enough for Gentry to reach out for the specimen. "Oh, that's *Gongora,*" he declares, as casually as if he has just spotted an old acquaintance on a city street, a kind of orchid, more delicate than a string of Chinese lanterns, with eighteen little yellow-and-red-speckled flowers hanging in a chain. It is one more small clue to what this forest on a coastal ridge in western Ecuador is all about.

It is also a slightly extreme instance of what people mean by rapid biological assessment, a hot idea in the environmental community just now. Gentry is here on behalf of Conservation International, a Washington, D.C.–based group, which conceived its Rapid Assessment Program (or RAP) to fill a widely acknowledged void in the movement to save the rain forests: For all the attention the issue has received in recent years, nobody really knows what's out there. In allocating their limited resources, conservationists have little scientific basis for choosing one forest over another. A rain forest is more likely to get protected, if it gets protected at all, because nobody else wants it right now, or because somebody's uncle used to hunt there and remembers it fondly, than because of its biological value. To fill in the blank spaces on the map, RAP employs the most knowledgeable tropical field scientists in the world, people who normally spend years studying one small patch of forest, and sends them instead on quick sweeps to assess vast threatened regions.

Their January 1991 mission in Ecuador is typical. In little more than a month, four U.S. and three Ecuadorian scientists are to survey the entire four-hundred-kilometer coastal range between Guayaquil and Esmeraldas, and determine what areas have survived deforestation, which plant and vertebrate species still live there, whether a significant number of them are endemic (found nowhere else), and what the priorities should be for conservation. The Conservation International team, which conducted its first survey in 1990 in Bolivia, was assembled with the aim of making a dozen such sweeps in different countries over three years.

It is biology on the quick, and it sounds at first like anathema to traditional science, which is oriented toward collecting specimens and filing them away in the back rooms of the world's great museums, and to producing tentative and methodically demonstrated conclusions about biological minutiae, to be carefully reviewed by other researchers and published in obscure professional journals five years hence. RAP aims instead to see the big picture now and act on it, while there is still time. The program has been called "an ecological SWAT team" and a "Biosquad," terms with an infelicitous suggestion of do-good gringos in jackboots. Conservation International's approach, on the contrary, is to turn its findings over to the country in question, and then raise funds and apply judicious pressure to help local groups develop a practical environmental response.

The four permanent members of the RAP team are extraordinary. Gentry, age forty-five, is a senior curator at the Missouri Botanical Garden, and is often introduced at speaking engagements with the flat assertion that he knows more tropical plants than anyone who ever lived. He is a tall, gangly figure who dresses like a motorcycle mechanic, in a grimy T-shirt and blue jeans that are tattered at the cuffs from chronic tree-climbing. His eyes are bloodshot from staying up till 2:00 A.M. pressing botanical specimens; they give the unsettling but entirely accurate impression that he will do anything to get to a plant. Bitten once by a venomous pit viper four hours by boat from the nearest thatched-hut village, his first reaction was annoyance that he would be losing field time. The snake died; Gentry recovered.

Ted Parker, a thirty-seven-year-old research associate at Louisiana State University's Museum of Natural Science, is Gentry's equal in ornithology. He can identify most of the four thousand bird species of the New World tropics by their songs alone—for example, picking out the subtle differences that distinguish each of the three hundred seventy kinds of tyrant flycatcher or the three hundred or so different antbirds. "A lot of people say they sound the same," he says. "But to me, every note is distinctive." This phenomenon tantalizes and also strikes fear into the birding world; Parker first made his name when he visited a heavily studied site in Peru and promptly identified fifteen species his colleagues hadn't noticed in a dozen years of research.

The RAP mammalogist is Louise H. Emmons, age forty-six, a Smithsonian research associate. Her colleagues like to recount how a radio-collared jaguar once stalked her in a Peruvian rain forest at night as she walked to camp, a tracking receiver clicking furiously in one hand. But Emmons disdains the story for its emphasis on charismatic megafauna; she prefers to explore how unglamorous squirrel species partition their habitat in the rain forest of Gabon, or how tree shrews live in Borneo, from which she has just flown in. Her *Neotropical Rainforest Mammals* is the only comprehensive field guide for distinguishing among the Western Hemisphere's eighteen genera of rain-forest rat, for example, or the thirteen species of porcupine.

The fourth member of the team is Robin Foster, a forty-six-year-old plant ecologist with the Smithsonian Tropical Research Institute (STRI) in Panama and the Field Museum of Natural History in Chicago. Foster is best known for the most thorough long-term study of rain-forest dynamics ever attempted, in which he and a hundred assistants tagged, identified, mapped, and measured—and now periodically revisit—a quarter million plants on a fifty-hectare plot at STRI's Barro Colorado Island station. He is adept at looking at a forest from a satellite image, an airplane overflight, or a road two miles away, and judging what grows there and whether it is old-growth forest. If a paramilitary metaphor is inevitable, then he is, among other duties, the reconnaissance man for what is, in effect, a search-and-preserve mission.

In early February 1991, near the end of the longest drought in Ecuador's history, the countryside around Portoviejo is a dust bowl. Every hillside is denuded, and fissured by erosion. Here and there, commercially worthless ceiba trees, the last remnants of the old forest, lie on the bare fields where farmers have cut them down and burned their branches. The huge lime-green trunks with their sinuous ridges look like slaughtered elephants. "I've never seen it this bad," someone remarks. But someone else recalls the deforested hills of Greece and the overgrazed landscape of the American West—comparisons that are flawed only because of Ecuador's superior biological wealth. With an area the size of Nevada, it possesses more plant and bird species than all of North America.

Parker and Foster are scouting ahead for their next study site, while the rest of the team finishes up in the south, at Machalilla National Park. This region, the densely populated western half of Ecuador between the Andes and the Pacific, has already lost 95 percent of its forest, most of it in the last twenty years. But Foster knows from an overflight in December that patches remain on the ridgetops of the coastal mountain range, which the flood tide of human population may not reach for another year or two.

Practical considerations might suggest writing off these western remnants as unprotectable. The immediate needs of the many small farmers here take political precedence over conservation, even when the *campesinos* themselves recognize that cutting down the forests to feed their families this year may mean having nothing to eat or drink five years from now. At Machalilla, the group found hunters even inside the national park; loggers had cut down most of the trees, and livestock grazed on the hilltops.

RAP has chosen this unpromising terrain because the isolated ridgetops may contain the last pockets of endemic species that have otherwise disappeared from western Ecuador. Apart from their biological interest, these pockets may also prove of long-term human value. They may include commercially valuable species—a seed stock for developing some more sustainable way of life when the short-term advantages of clear-cut logging, uncontrolled hunting, and subsistence agriculture inevitably fail.

But on the road north to Pedernales this hope dims. With Parker at the wheel, the car bucks and lurches for hours across rough dirt roads, past vegetation that has been nibbled to the root crowns. The endless goat and cattle trails cut a sagging latticework pattern into the powdery face of every hill. The mangrove swamps have been flattened and converted to rectangular shrimp farming ponds. In the dusty little down of Jama, the owner of an open-fronted corner store explains that he has seen the forests disappear in his lifetime because of agrarian reform. Beginning in 1964, reform made undeveloped land available for the taking; the claim-holder was obliged only to clear the land and put it into production quickly, or risk government confiscation. The store owner indicates the parched, ovenlike atmosphere with a gesture of resignation, and adds that since the forests disappeared the rains have not come. Parker and Foster push on to the north. In fourteen hours of driving, they encounter only one sizable piece of forest, on Cerro Pata de Pájaro, or Birdfoot Mountain. It is smaller than Central Park in Manhattan.

Parker resists the suggestion that saving such a meager remnant would be futile. "If that's so," he says, "you have to tell tens of thousands of people working for the Nature Conservancy back home that they're wasting their time, buying one or two acres at a time." If Birdfoot Mountain were an oceanic island, he adds, biologists would be all over it in search of endemic species. But when the entire team regroups, his frustration with the long threadbare stretches of coast rushes to the surface: "It's fucking *gone*. Believe me. There are patches of two or four or five hectares you can work in. But it has no conservation future."

"For birds," Foster agrees mildly. "But for plants?" They decide to head farther north, where the climate is more humid and the forest less hospitable to human invasion.

Regrouping also means a brief opportunity to recover. Emmons, who picked up bubonic plague on the Bolivia trip, is now fighting headaches and fever of unknown origin. A rich medical history appears to be one of the chief rewards of field research, along with uncertain pay, troubled marriages, and frequent arrests (Gentry has been jailed fifteen times, usually for seeking plants in the wrong place at the wrong time). Among them they have endured malaria, encephalitis, filariasis, leishmaniasis ("The best thing is, it eats your cartilage"), hepatitis, botflies ("Chew up a cigarette and stick it on the point of entry with a Band-Aid"), arboviruses, giardiases, worms, amoebas, and assorted bites and stings. Emmons shrugs it all off. She says the biggest danger in the rain forest is that a branch will drop on your head, and she notes that in Africa an overdose of antimalarial drugs is a common means of suicide. Parker, still despondent, remarks that if they don't find a forest soon, mass suicide might be in order. What Parker really needs, Foster suggests afterward, is a medicinal dose of eighty bird species in song, applied twice daily. What all of them need is the consolation of rain forest.

Heading north that night, Gentry finds himself stuck at a barricade of dirt and downed trees across the road, with flames rising here and there. Someone is playing music on loudspeakers, and the local villagers mill about yelling "*La Huelga! Viva!*" It is a provincial strike, a relatively common occurrence in Ecuador. Gentry, who does not like to lose field time, guns the engine of his Land Rover and says, "I think I can take it." This is entirely in character. One of the group's other vehicles, an Isuzu Trooper, has a large dent in the back. Parker was driving it up a muddy slope earlier in the trip, and struggling—until Gentry banged the Land Rover into his rear end and pushed him to the top. Another time, Gentry discovered that a vehicle he had donated to a South American university was being used by administrators instead of field researchers. He stole it back and delivered it to a more sensible university in another country.

Parrots

This time, someone persuades Gentry to be patient with "*la huelga,*" and they sleep three abreast on top of the Land Rover at the side of the road. In the morning, it turns out that there were six more barricades beyond the first, including one at which a couple of tractor trailers have had their tires burned out from under them.

The RAP team's luck doesn't change till three days later, south of Esmeraldas, when Parker stops the car, looks out at the blue ridges piled up in the distance, and says, "There it is." By good (or not so good) fortune, they find a new logging road into the heart of the forest, along the red clay of a ridgetop. As he bounces in his seat, Parker mutters to himself like someone coming out of image deprivation: "This is forest. There are birds here. There are more plants than you know what to do with." Foster, more circumspect, studies the treetops. "That's good," he comments. "Lots of virolas. Same family as the American nutmeg. They're taking stuff out. But they're high-grading. There's lots of other stuff still intact." Further in, he adds, "A lot of things I've never seen before."

"I gotta listen," says Parker. "Just to see. Please." He stops in the road. "Oh, man. This is gonna be great. Oropendolas. All the tanagers." Later, on foot, he says, "That accelerating sort of staccato is a blue-tailed trogon. I've never heard it before." How, then, does he know what he's hearing? "Because it's the only neotropical trogon I haven't heard." A moment later, by way of proof, a big bird with a metallic blue tail lights on a nearby branch. But Parker is already off among the songs of rufous pihas and immaculate antbirds, and among the flocks of big parrots raucously calling to one another across the road. Foster finds the logger at work and makes arrangements for the RAP team to return the next day. Noting the softness of the road on the drive out, Parker says, "Just hope it doesn't rain tonight."

The rainy season of course begins before dawn, torrentially. With the road now passable only by a long hike through ankle-deep mud, the true romance of field research begins to impress itself on everyone. During the next four days of continual rain, the RAP team is all over the forest, sodden, but methodical about taking its measure. Luis Albuja, the leading authority on Ecuador's 125 bat species, rigs his mist nets. Emmons baits fifty box traps for small mammals and hides them in tree cavities and beside fallen logs. Gentry sets up his transect to make a detailed sampling of a tenth of a hectare of forest, before the logging road can claim it. Foster heads out into *bosque puro,* the intact forest, on day-long hikes to assess the area's large-scale ecology.

The team is camped with the logging-road crew, and the combined traffic produces a quagmire. People struggle to remain upright and in forward motion, walking gingerly, as if they have just had casts removed from both legs. Having slogged through the mud all day, many of the researchers strap on headlamps and hike out again at night. Albuja visits his nets to delicately untangle fruit-eating bats with needlelike teeth and leafy noses. Emmons walks alone in the underbrush, dead slow, listening for mammals and looking for their telltale eyeshine. A pair of herpetologists seeks tree frogs along the sides of the trails. Headlamps move across the foliage in robotic slow motion.

Under a tarp at the camp, by the phosphorescent white light of a lantern, Gentry is pressing plants, which means that he is slapping leaves, fruits, flowers, even a spiky, coral-like palm tree inflorescence between pages of a local newspaper, *El Telégrafo,* wherein an op-ed columnist inveighs against *"Ecologismo exagerado."* As he works, Gentry is talking about a site he and a colleague once studied in central Ecuador. On a ridge called Centinela, near a small, privately owned reserve called Río Palenque, they identified nine hundred plant species, including about a hundred found nowhere else. Then farmers clear-cut the ridge, reducing it to bananas and perhaps thirty species of weeds. He holds up a red, podlike flower, with a fluted undersurface like the belly of a blue whale. "This is *Gasteranthus,*" he says. "Stomach flower. You can see why it was called that. And I think it's a new species. Anyway, I've never seen it. There are twenty-four species, and six of them are endemic to Centinela. So they're presumably extinct."

He speaks without bitterness or other emotion, except a kind of detached, driving intellectual curiosity. As the group had theorized, plants that have disappeared from Centinela and other devastated areas of western Ecuador are turning up here, mixed in with plants from the rich Choco forest of western Colombia, and with new species found nowhere else. The character of this forest is beginning to take shape in the group's collective mind.

Among other things, Gentry has seen several Río Palenque mahogany trees, a species often cited as an instance of the mindlessness of unchecked deforestation. The tree, a mainstay for pre-Columbian builders, was the most commercially valuable timber in northwestern Ecuador until about 1970, when loggers wiped it out. About a dozen trees of the species are known to survive, all at the Río Palenque reserve, which may be too small for long-term stability. Discovery of additional trees on a ridge near Esmeraldas would be important news, except that old habits die hard: The logger intends to harvest all this lumber and sell it, for about $3 a tree, to produce plywood. Incidentally, says Gentry, the stump being used to hold up the RAP team's dinner bench is a new genus.

As Gentry is about to press a flower from a tree of the coffee family, Emmons stops by to visit. The atmosphere in this group is one of eager, almost avaricious, swapping of information, with a moderate undercurrent of friction. They are all headstrong, accustomed to working alone, often going from year to year on grants. Emmons and Gentry have worked together intermittently (once on a rich forest later cut down to make cardboard boxes), and an edgy note of affection and intellectual one-upmanship characterizes their relationship. Gentry has suggested, for example, that a lizard he saw might be a chameleon because it changed color. "I'm sorry," Emmons snaps, "that doesn't make it a chameleon." Anyone but a botanist would know South America has no chameleons.

Now Emmons looks at the flower Gentry is handling, with its eight-inch-long tube leading from the petals to the pocket of nectar at the bottom. "Isn't that the one Darwin looked at and said there had to be a moth with a tongue that long to pollinate it?" she asks.

Gentry grins. "And when they found the moth," he says, "they named it *praedicta*." They relish this piece of lore for a moment. Then Gentry says, "It's actually not. In fact, the one Darwin saw was a Madagascar orchid, *Angraecum sesquipedale*. And *praedicta* is a moth subspecies."

Foster heads out one morning on what he describes as a quick tour to gather some specimens. He is the least aggressive member of the team, with a habit of muttering small, self-deprecating jokes, followed by a soft titter. The personality, along with the information that he is operating on only a third of a liver due to various tropical diseases, can be dangerously disarming for anyone trying to tag along. "See you at the bottom," he says, chuckling, as he heads down a landslip where the footing is slick, the angle near-vertical, and the only available handholds covered with thorns or rooted so loosely they tear out as he plunges past.

Later, about to climb a waterfall that has gotten in his way, he turns and remarks that as a child he was influenced by the movie *Lorna Doone,* in which a young woman lived with a mean-spirited family on a peak accessible only by way of a waterfall. Foster spent his teenage years seeking her on waterfalls near his parents' home in Vermont—what animal behaviorists might call displacement activity. "At the time, she seemed worth climbing waterfalls for," he says, and then laughs and disappears up the wet rocks with a Santa Claus sack of plant specimens on one shoulder.

As he walks, Foster speculates about the anomalies of this forest, which he describes unequivocally as the most interesting he has ever seen. It puzzles him, for example, that the forest contains almost no large palm trees, though these trees are abundant just to the east. Wind-dispersed tree species are also absent, and while tree-climbing vines and epiphytes are spectacularly diverse, the family called Bignoniaceae, which would normally predominate in a wet forest, is missing, as is the tree family Leguminosae, also normally abundant. Despite a tendency in the public mind to lump all rain forests together, each is distinctly different. This one, says Foster, is "a weird combination" of plants associated with both rich and poor soils.

The botanical puzzle also provides clues to the whereabouts of many animals. An abundance of fig trees may partly explain the noisy flocks of parrots; Parker has spotted four species in a single tree. But mammals are scarce. Emmons's traps have come up empty, and her walks have yielded almost nothing: howler monkeys, an opossum, kinkajous, squirrels, and some tracks of armadillos and pacas. At breakfast and dinner, or when they meet in the woods, the researchers trade half-formed ideas: Does the scarcity of rodents correspond to the lack of wind-dispersed plants, with their edible seeds? Shouldn't the virolas here attract something other than toucans?

"The virolas have this oily, fleshy part around the seed," says Foster. "In Panama, spider monkeys love these things. But there aren't any spider monkeys here. Agoutis eat them, too, but there are few agoutis here. Have they all been shot? It's hard to believe." Emmons speculates that the long drought may be partly to blame.

Their talk veers from new discoveries and inchoate theories to all the small oddities that make a rain forest so interesting. Parker explains the ability of some wren pairs, once bonded, to sing antiphonal duets—one song with two birds indistinguishably trading alternate notes—but only if they are perched less than twenty inches apart on a branch. John Carr, a herpetologist, shows off a thumbnail-size frog with a glistening knot of tadpoles on its back. Whereas other frogs abandon unhatched eggs to float on the water surface (you can see them everywhere in puddles formed by yesterday's footprints), the male of this species apparently tends its eggs in the leaf litter, and then carries the tadpoles to water. A conversation about how destructive local hunters can be for spider monkeys, which breed only once every four years, nosedives into a discussion of the culinary merits of monkey species served to them at past logging camps. "The worst thing," says Gentry, "is when you find a scrawny arm sticking out of your stew." On the other hand, howler monkey doesn't taste too bad.

A running theme of every conversation is how to save some remnant of this idiosyncratic forest, and of western Ecuador, from the stewpot. Emmons urges practicality, persuading the government to protect at least one ridge at Machalilla, if not the entire national park. Parker vacillates between pushing for someone to buy up forested land around Esmeraldas, or simply walking away, to concentrate instead on Machalilla and an ecological reserve to the northeast called Cotacachi-Cayapas, now threatened by loggers, colonists, and a proposed highway.

For Parker, the adjustment to Ecuadorian-style deforestation is plainly difficult. He is accustomed to working in Peru, where colonists start at the roadway and spread out systematically on either side. But in Ecuador, settlers are everywhere, leapfrogging two hours into the forest, to stake a claim where a road might reach them five years from now. He is accustomed to Bolivia, where the population is smaller and the conservation movement far more effective. In Bolivia, it is entirely possible that the RAP team's discovery of a biological mother lode at Alto Madidi, near the Peruvian border, may lead to the creation of a park on the scale of Yellowstone. (In two weeks there, Parker found 403 bird species, and Gentry logged 204 plant species in a tenth of a hectare.) So when Parker considers the diversity of birds and plants still hanging on here in Ecuador, it seems modest to conjecture that a reserve of about twenty thousand hectares would be adequate protection. With small plots selling for $5 a hectare, you might be able to pick up the whole thing for $75,000, the salary of a middle manager at some oil company or environmental group back home. Foster gently informs him that the area south of Esmeraldas may have only a thousand hectares of unbroken forest left.

Parker is the leader of the RAP team and an immensely likable character, enthusiastic, informal, outspoken. He has thick, dark hair, finger-combed off his forehead, and an untrimmed beard under his jaw. His broad, reddened cheeks frequently lift up in an easy grin, making his eyes even narrower. He has been obsessed with birdsong since he was six or seven years old, and found to his delight that it enabled him to impress his elders on nature tours around Lancaster, Pennsylvania, where he grew up. He worked on bird notes in class and once, in high school, simply stood up and wandered out of the building, explaining afterward that he had heard birds. His career has defied convention, especially academically. As a freshman at the University of Arizona, he took golf for his physical education credit, golf being something he could have in common with his lawyer father (even if Parker insisted on bringing his bird binoculars out on the links). Unfortunately, it was the only course he passed. Otherwise, he thought about birds (and pickup basketball, a lesser obsession) and spent far too much of his time on long birding road trips into Central America. When he eventually graduated, he went to work conducting ornithological surveys as an unpaid associate at the Louisiana State University Museum of Zoology, and leading more than fifty natural history tours in Latin America for Victor Emmanuel Nature Tours. Along the way, he has recorded more than ten thousand birdsong tapes for Cornell University's Library of Natural Sounds. His travels ultimately gave Parker the idea for RAP and also led him to point the group first to the unknown riches of Alto Madidi.

Sometime after the last crested owl has called, but before the rufous motmot begins the dawn chorus, Parker starts his workday, unconsciously filtering out the sounds of raindrops drizzling from leaf to leaf, of monkeys howling in the distance, of assorted nonavian creatures whirring, buzzing, and croaking. He stands under a faded umbrella, with binoculars around his neck and his microcassette recorder up to his mouth. His khaki pants are tucked into the tops of clean white socks and his sneakers are sunk to their high-tops in mud.

"Amost everything you hear is ochre-breasted tanagers," he whispers. "It's very difficult to hear because they have such a variety of song, and it's constant. Not just melodious things, but scratchy songs, too." He cranes his head up sharply and says, "*SHHH!*" And then: "These birds are just so fucking loud." He stands and seems to do nothing. The intensity of his listening becomes apparent only afterward when one hears what he has murmured hesitantly into his recorder: "Scarlet-rumped cacique . . . a fasciated antshrike with a pair of barbets singing nearby, probing at a cluster of leaves . . . and feeding on fruit . . . a pair of *Heterospingus* . . . two more pairs of *Myrmeciza immaculata* counter-singing. *Dysithamnus puncticeps* chorus, male and female. . . ."

After a while, he points to a broad swath of roadside forest hacked down yesterday by a crew of twenty men. "Look at this. This was my best area. I'm trying to think of any place in the world where people work all day in the rain without stopping, with machines. It was eerie to hear these chainsaws in the fog.

"Chestnut-backed antbird. I wonder. They've had a big part of their territory cleared and they're still singing." It is a plaintive two-note, a voice saying "Come here." "That soft wolf whistle is a barred puffbird." A languid sound, the whistle of a man already lying in another woman's arms. "Here comes a flock of blue-headed parrots, 1-2-3-4-5-6-7-8." He pauses momentarily. "You hear that cricketlike sound? That's a short-tailed pygmy tyrant, one of the smallest passerines. Most people overlook it because it sounds like an insect." He stops at a tree that is dancing with birds, records them, and then plays the tape back, enticing them to stay in the area and peek curiously from the foliage. "That sounds like an ochraceous attila," he says. Then his head whips around: "There it goes! *LAND!*" It does, and in a few minutes, he picks up thirty-odd species.

His list for the four days at this site will end up 195 species long, a healthy total for deep forest; Gentry will log 120 plant species in his transect. From the entire monthlong trip, the RAP team will conclude that only a fraction of western Ecuador remains forested (the Ecuadorian forest service, DINAF, says 24 percent is forested—but bases this on a 1977 map), that this remnant has both biological and long-term commercial value for Ecuador, and that it is in immediate danger of being destroyed. The RAP report will also add to the growing evidence that national parks, which supposedly protect 11 percent of Ecuador, are meaningless boundaries on the map.

A little after 8:00 A.M., between the calling of a russet antshrike and the beelike passing of a barb-throat hummingbird, a chainsaw starts up. "This land is going to be gone in a few years and the people of Ecuador are going to have to do something else to survive," Parker says. "Nobody will think about what's lost until fifty years, a hundred years from now, when people here are going to say, 'Why didn't we just buy a little?' The same thing happened in the United States. Why did we cut the Singer Tract in Louisiana, where the last ivory-billed woodpeckers were? It was in the 1940s. We cut it down for railroad ties." He grimaces. "Why not have a few thousand hectares of pure forest preserved? If we suggest these things, maybe some Ecuadorians will come forward and start to fight for places like this."

At the end of the logging road, the bulldozer comes clanking and roaring to life. Even in the rain, it moves forward two hundred or three hundred meters a day, and it has arrived now at a massive tree in the middle of the old trail. *Coussapoa villosa,* according to Gentry; once a strangler vine, now a tree ten feet across at the buttressed base. It started as a seed dropped in the top of a tree here by a bird. After growing for a time as an epiphyte it sent down aerial roots in search of the earth, and the tendril-like roots gradually twisted together and became a tree themselves, squeezing their host to death. Other plants now grow on every available surface, perhaps two dozen species in all—orchids, spiky bromeliads, ferns, the showy red fruit spike of an anthurium, the broad elephant-ear leaves of a philodendron.

The bulldozer deftly undercuts the left side of the tree, and then moves to the right, where the strangler leans heavily on a tree-size aerial root. The operator's hands move skillfully among the levers, going from forward to reverse, adjusting the angle of the blade, accelerating into the root. It is hard not to admire his ability. He earns about $5 a day, possibly with funeral expenses thrown in if the tree drops on top of him, as it gives every indication of doing. Foster points out that Americans once built a mythology around this sort of thing—pioneers risking their lives to break the earth for their families. It is a vast leap from that mythology to an understanding of its consequences—maybe an impossible leap for people who still aspire to the prosperity North Americans take for granted. Roots crack apart, and after thirty passes, the tree shudders and begins to give way. The operator glances up and sees the strangler starting to fall. He throws his bulldozer in reverse, and the fabric of plant life begins to rip apart in front of him. Vines separate, branches split, leaves shower down. The strangler arcs past with a slow whoosh, hitting other trees nearby and dragging them down with it, until the whole canopy seems to come crashing to earth. Light floods into the ensuing silence.

"*Muy grande,*" a spectator comments.

The operator reaches one hand up to the open sky. "Look at all the sunlight!" he says. He contemplates his work for a moment with satisfaction. Then he throws the bulldozer into gear again, and pushes on into what is left of the rain forest.

Postscript

In September 1995, the government of Bolivia created the 4.46-million-acre Madidi National Park, protecting an area the size of

New Jersey that had come to international attention as a result of the first RAP mission five years earlier. Here, Parker had found sixteen species of parrot, an arctic tern never before recorded in the South American interior, and nine bird species unknown in Bolivia, from a scarlet hooded barbet to a casqued oropendola. He suggested that careful study of just one swath of the terrain, which is in the upper Amazon basin, on the eastern slopes of the Andes Mountains, could yield more than a thousand species of birds, "an amazing 11 percent of all bird species on earth." A year later, on the adjacent slope of the Andes, Peru created the 802,750-acre Bahuaja-Sonene National Park.

Neither Parker nor Gentry lived to see this success.

In August 1993, they were back in Ecuador, and a local conservationist proposed an overflight of a dry forest on the coast. Parker didn't want to go. In the RAP report on their 1991 mission through this area, he had written: "By the time this document is published, much of the forest that we saw during our travels through western Ecuador will have been destroyed." But the Ecuadorian conservationist was unwilling to abandon the area. Parker and Gentry dozed through much of the flight, droning over bare, ruined landscape. They crossed into upland forest, and the pilot of their small plane seemed to be lost. He turned, flying low, and went into a cloud. On the other side was a mountain.

Parker, who was age forty, and Gentry, forty-eight, were both trapped in the wreckage and died there the next day. Recovering a few months afterward, a survivor, Parker's fiancée, told a *Washington Post* reporter, "I think it was a good place for them to die. It was beautiful forest, and they were very happy." She hesitated, then softly added: "Lots of birds."

Among conservationists, word spread quickly about the loss of these extraordinary men. One ornithologist wrote, "I feel like all the libraries in the world just burned up."

Sleeping with
Snapping Turtles

Almost everybody has a snapping turtle story to tell, and the peculiar thing is that they always have nasty endings. By and large these tales fall into two categories. In the first, the snapper displays its daunting ferocity. My brother, for instance, once saw a snapper crossing a highway. As he got out of the car to help, a tractor trailer roared past. The snapper could be seen between the wheels, striking into the air with its neck extended and its hooked beak open as if to snare the thing by its axles. My brother got back into his car and drove quietly away.

In the second category, the snapper meets its brutal end at the hands of hunters or bird lovers outraged by its supposed incursions on the duckling population, or from adolescents who take its hissing malevolence as an affront, or from people who merely want snapper soup. These stories all have to do with the snapper's implacable and unsettling hold on life. Thus a man living in a town not far from me killed a snapper for soup and buried its head in his yard. His dog dug it up two days later, he says, and it bit the dog on the nose.

No less an authority than Archie Carr, the turtle scientist, grew up in what he terms "more or less sustained awe" at one such snapping turtle story, from the popular boy's novel *Rolf in the Woods,* by Ernest Thompson Seton. Here Indian Quonab stalks Bosikado, the hundred-pound "devil of the lake," who has driven away the ducks and robbed the fish lines. Ahab-like, the Indian finally meets his "goggle-eyed monster" in deep, muddy water. Bosikado locks his beak on Quonab's left arm, Quonab plants his hatchet in Bosikado's back, and the two of them fight to the death. Or something like death. Quonab emerges from the roiling water with the turtle's decapitated head still clamped on his forearm, and for an hour or more the headless body tries to drag itself back to its lake. "Mine enemy was mighty," Quonab sings, "but I went into his country and made him afraid."

Though we seldom see them, most of us have the uneasy and entirely accurate suspicion that snapping turtles are out there lurking in our swimming holes and ornamental ponds. Even for people who ought to know better, it is an irresistible leap of the imagination to the idea that snapping turtles are underwater dinosaurs, lake-bottom *Jaws,* ready at any moment to grab a toe or forearm and drag us down. In my neighborhood, on the Connecticut River, lives an unusual character whose home has simultaneously sheltered thirteen pitbull terriers, a raccoon, two hundred copperhead snakes, a bamboo viper, a boa constrictor, and enough rats and mice to keep everyone from going hungry. He used to water-ski on the river, he told me recently, until he saw the size of the snapping turtles living in it.

Now I was inclined to take all this skeptically, in part because few animals ever turn out to be as big or as bad as people like to imagine, and also because I still swim in the Connecticut River (along with my children, who weigh less than some of the snappers I'd heard about). Then one summer, I saw a newspaper photograph of a bearded man in a pith helmet holding the "biggest wild snapping turtle on record"—no Bosikado, perhaps, but big enough: sixty-seven pounds, with a 19.8-inch shell, and an overall length, neck extended, of "well over three feet." Its captor, identified as "reptile hunter John J. Rogers," had reportedly done something hardly anyone ever thinks of doing with a snapper. He set it free in the Massachusetts pond where he'd found it.

At that point, I'd been looking for John Rogers off and on for the better part of a year, and I'd heard almost as many stories about him as about snapping turtles. Across his range, which seemed to include New Jersey, where he was born, New York and Massachusetts, where he did most of his turtling, and Maine, where he had a camp for fur-trapping, Rogers had acquired an almost legendary reputation for his ability to catch snappers.

The methods other outdoorsmen practice on snapping turtles tend to be cumbersome. Some of them run lines of baited hooks off of riverbank saplings, or use empty Clorox bottles as floats for "jugging" snappers. Some of them set out big, barrel-shaped traps made of fencing, with dead fish for bait. A few turtle hunters go out in winter, find a likely spring hole, and probe around with a metal pole in the mud till it goes *thunk* on the shell of a hibernating snapper. Others flop down in the mud and feel around blindly under riverbanks till they find a snapper to pull out, preferably tail-first—a method known as "noodling," the practitioners of which can be numbered even on the fingers of *their* hands. Rogers reportedly disdained all these methods. It was said that he could actually call the snappers to his boat by a technique unknown to science, and that he did it well enough to make his living as a turtle hunter.

Unfortunately, Rogers was also said to be intensely secretive. I had a phone number for a friendly couple living near his Maine camp, and every few months I left a message with them telling Rogers that I intended to write an article about snapping turtles

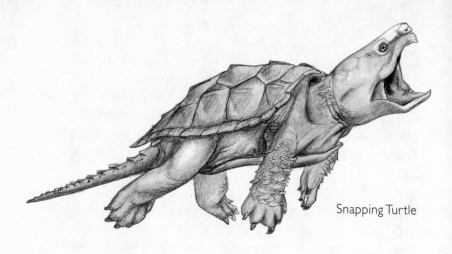

Snapping Turtle

and asking if he would call. "He told us he'd be back at the end of June," the woman said, apologetically. "But he didn't say what year." At various times he was reported to be over in New York turtling, or out in Ohio selling the week's catch, or down in Florida hunting alligators. "He comes and goes like a butterfly," a friend of his told me. "Find a pond and keep watching long enough, he'll probably turn up." But if he did, he wouldn't talk about his work. And he added, "He had three thousand pounds in his pickup truck not long ago and he wanted me to go look at 'em. I didn't want to look at that many snapping turtles. I don't much like the damned critters."

And who, after all, does? To the average eye, the snapping turtle isn't just mean, but ugly. It has a long, ridged, alligator tail (a "knife-edged war club" on Bosikado). Its stocky, armor-plated legs end in sharp yellow claws, like overgrown fingernails. On its underside, the cross-shaped plastron exposes loose skin with the mottled, pinkish-yellow complexion of plucked chicken. Leeches often attach themselves here and on the neck, which is cowled with warty flesh. The eyes, set well forward, are the same mottled black and brown as the skin around them, and the black pupils are inscrutable. The head comes to a point in two close-set nostrils, like pinholes; it rises and falls subtly with the effort of smelling, and this motion gives the snapper the appearance of a sumo wrestler, laboring under physical massiveness and barely restrained spiritual malevolence. Its mouth, though toothless, has a sharp bony ridge, forming a dyspeptic smile to the hooked tip. When the turtle is angry, the mouth gapes open, pink and hissing. It can also strike, according to Archie Carr, "with the speed and power of a rattlesnake."

It's probably just as well that most people never see snapping turtles or experience the resulting "thrill of unnervement," to borrow Ernest Thompson Seton's apt phrase. Even so, the snapper is ubiquitous in its range—east of the Rocky Mountains in this country, and from New Brunswick to Ecuador. Snapping turtles spend much of their time underwater, lumbering along the bottom looking for food, or waiting for something edible to pass within striking distance. And despite their storied appetite for truck axles and dogs' noses, they are shy of human contact. Most sightings occur in June, when the female travels as much as a half-mile overland in search of a suitable place to lay her eggs. According to Eric Kiviat, an ecologist at Bard College, young snappers may also migrate overland. "It's well known that if you dig a new pond in the eastern United States, they'll appear," he told me. But the larger snappers he surveyed tended to be recaptured within a hundred yards of where he'd tagged them, and one ponderous forty-two-pound male turned up four years in a row in the same quarter-acre pool.

This sedentary, underwater way of life means not even scientists know much about snapping turtles. On certain questions, Kiviat happily deferred to a highly successful trapper who had worked in the area in the 1960s and 1970s, an accurate and astute observer, whose ideas about snappers merited scientific study. Kiviat's research papers from the period referred to him as J. Rodziewicz, and Kiviat remembered "Jasz" (pronounced "Yosh") but said he also sometimes went under the name John

Rogers. Unfortunately, he had not seen or heard from him in several years.

John Rogers grew up on the Pequest River, a tributary of the Delaware, in the 1940s, in a Polish Catholic family that survived on trucking, small farming, and whatever it could forage from the rural New Jersey countryside. As a Christmas treat, the family ate sucker cakes made of ground mullet, a trash fish that was far sweeter than hatchery trout. Rogers used to go out with a flashlight to catch them, cutting a hole in the ice on a local millpond, pounding on the ice to make the mullet bolt past the hole, and then hooking them with a three-pronged gaff. In the summers, his father put him to work on the farm at a dollar a day, and he spent his spare time trapping and studying the millpond. The girl next door rowed the boat, while Rogers hung over the side. At the age of nine, he caught his first snapping turtle. He put it on a family truck running between Boston and the fish stores of eastern Pennsylvania. The truck came back with a day's wages, $1, in payment. Rogers began to go out on the ponds and irrigation ditches with his mullet gaff, and he caught twenty-eight snappers that summer.

"So that was the beginning," he told me one night last November, when I'd given up on the phone and simply driven to his Maine camp on the chance of finding him there.

The camp was a one-room log cabin Rogers had built, which he was sharing then with a sometime partner, a fractious eighty-six-year-old named Norman Nooney. It was the height of the fur-trapping season in Maine, and Rogers was off working when I arrived. As we waited for him to show up, Nooney sat muttering imprecations at a small black-and-white television set tuned to a *Bob Newhart* episode he deemed humorless, and, in between, talking about snapping turtles. He and Rogers met in the 1960s on the Connecticut River. At the time, the older man was waging a solitary crusade against the snapping turtle, having taken 2,161 of them in three seasons, apparently to avenge the loss of some goslings. Rogers paddled up one day, and when Nooney asked, "Who are *you?*" Rogers said, "I'm a turtle hunter"—the beginning of a beautiful friendship. Nooney told me that Rogers was six feet, five inches tall, had hands like hams, could loft a ninety-pound canoe the way ordinary people lift a feather pillow, and would not talk, all of which, excepting the last point, turned out to be true.

"I like snapping turtles," Rogers said, shortly after his arrival. Their shells, and faces, and mannerisms had individuality for him; he could look at a load of them banging around in the back of his truck and tell where each had come from. He liked them well enough to give his record-breaking turtle a name, Dave Butz, after the thick-necked Washington Redskins defensive lineman. "This turtle was rather nonchalant and lackadaisical about me, after the first few snaps," he said. "I came to like him personally." Rogers had caught Butz before in the same cove, and he had also once caught a seventy-two-pounder, without realizing they were contenders for the record. "I always let the big ones go," he said. He had become "diabolically accurate" about guessing turtle poundage (a boast Kiviat had confirmed with a scale), and if they were under ten pounds or over fifty-three, they went back. They were his breeding stock and his future.

I had expected that a lifetime among the snapping turtles might have made Rogers snappish and unfriendly, the way a man and his dog or a husband and wife come to resemble each other. But in truth he was genial. If he was less enthusiastic about people than about snappers, he said, it was because he had reason to be cautious. At the one extreme were the people who wanted to limit trapping; "Save the Snapper" might not make it as a bumper-sticker slogan, but Massachusetts, Maine, and Rhode Island had lately begun to regulate his business. At the other extreme were the trappers who wanted to copy his methods and set themselves up in competition.

Trapping, he said, a little defensively, was not even remotely close to threatening the common snapping turtle population. (Its southern cousin, the alligator snapper, has required protection from overtrapping throughout its range, but it is slower-growing and less prolific.) "The snapping turtles right now are in the biggest population explosion of their history," he said, a development he attributed largely to the countless farm ponds built with government help from the 1930s through the 1950s. "There are so many more of them than anybody realizes," he said. "I went to a pond in Mexico, New York, and the farmer said, 'Snapping turtles? Catch all you want, but I'll tell you, I built that pond with my Daddy forty-two years ago and I've never seen one yet.' I took thirty-some snapping turtles out of there in an hour and a half."

Not even wildlife researchers appreciated the size of the snapper population, he said. A Massachusetts wildlife refuge once asked him to deal with a snapping turtle problem estimated at fifty individuals. "I got 1,284 snapping turtles averaging over twenty pounds. And I can guarantee you if I went in there now I could take a hundred in a morning.

"Next year is my twenty-sixth year," he said. "I've caught eighty-four thousand snappers now. The state I've hit the most is New York, and I haven't hit 95 percent of the water bodies."

The tricky part of the snapping turtle business wasn't supply, he said, but demand. Over the years, he had cultivated a network of buyers, including his first customer, who was now good for a thousand pounds a year. (A live turtle wholesales for fifty cents a pound, he said, and an experienced hand can kill and dress a thirty-five-pounder in under three minutes, yielding about a third of the live weight in edible meat.) The customers were spread out along the Mississippi River, and in odd pockets, among urban blacks, or in Catholic dioceses that once deemed snapping turtle an acceptable substitute for fish on Fridays. Rogers logged fifty-eight thousand miles last year supplying them. "I can sell twenty-three, twenty-four tons a year," he said. But one year he caught thirty-two tons and put himself out of business till the following June. "Once they have enough, trying to sell another is like trying to sell a road-kill groundhog."

As a result, he took a harsh line with competitors. "I've put eight different guys out of business in the last ten years," he said, an accomplishment that clearly pleased him. It was a simple matter of going to their favorite spots and, in effect, whistling up the snapping turtles while the competitor was still unloading his traps. "The Russian scorched-earth treatment," he said. In this fashion, Rogers had established himself as the king of the snapping turtles, and he regarded them with a proprietary and protective air. He had recently heard from a man living six hundred miles away who wanted permission to trap on the stream in front of his own house. Another man, to whom Rogers had ceded the turtling rights within fifty miles of West Winfield, New York, let him know that, on account of age, he would be needing only thirty miles henceforth. "Turtles are my people," Rogers said, and he made it clear that this was true roughly from Maine to Minnesota. "I don't trust these guys to know when they're overtaking 'em."

As he talked, Rogers had started working on the day's haul of mink and muskrat, deftly peeling off the skins and inverting them on tongues of wood to dry. The work stained his hands a slick berry brown, and the musk glands of the mink filled the air with their thick, spicy scent. I wasn't sure if he was talking to pass the time as he worked, or to make up for the phone messages he'd ignored. But in any case, he talked, and the tone of the conversation shifted from wariness about the wrongs he anticipated among people to a note of delight and contentment with the natural world, particularly as exemplified in the snapping turtle. His clear green eyes lighted up under a broad, unwrinkled brow, and the full beard and dark hair made him look closer to thirty than fifty. It was part of the family lore that his Rodziewicz forebears had been petty nobility in Poland, and he had the stature of a soldier-king. He was also plainly educated, and it turned out that he had graduated from New York University and studied European history at Columbia University (a temporary flight from the millpond) before quitting to become a benevolent despot among the snapping turtles.

It also became evident that he knew the land and his people in enormous and numerically precise, if quirky, detail. He knew the pond at the end of my street; he'd taken a thirty-eight-pounder there in the 1960s. He also knew Max Yasgur's farm near Woodstock, New York, and when he saw the film about the rock festival there he was curious to see people skinny-dipping in the very pond where, a little later in the summer of 1969, he took three thousand pounds of snapping turtles. "People swim with snapping turtles all the time," he said.

The reason they never saw them was that they had not trained their eyes. "I know where the turtles are, what they're doing, and when they're doing it," he said quietly. "I know when they're sleeping. If they're feeding, I know what they're feeding on. I can look at the rocks and trees and tell you how big the turtles are. I can look at the plants," Rogers said. He started shuffling through the lakes, ponds, and puddles in his mind, and sorting them into groups: "There's a lake on Cape Cod that looks like a lake outside Travers City, Michigan, and another outside Tomahawk, Wisconsin, and they all give big, black-shelled, white-to-tan-bellied snappers." In the Pocono Mountains, red rock meant blue-shell snappers; limestone kettles meant white. "I know a pond between Rochester and Syracuse. It's pure organic matter, just liquid mud, no rocks, no minerals. It has the smallest snapping turtles in the world. An old male eight and a half pounds is large." He reflected on that a moment. "I've got to take a course in geology. I'm going to take snappers from all the different areas and take sections of the shells and analyze them for trace minerals. To me, snappers are rocks that walk."

Were they also the sort of rocks who would come when called, I asked? Rogers said that he had witnessed all forms of snapping turtle behavior. He had seen them fighting in the spring, when the males tended to have white scars on the back of their necks from ducking and drowning one another. He had seen them mate. One rainy night he had watched 102 females laying their eggs in a field that had been plowed and disked. "Snapping turtles prefer to lay eggs on rainy nights," he said. Their nests are often raided by raccoons and other predators, and "in evolution, those turtles that lay eggs on rainy nights might have had more progeny, because the rain washes their scent away."

Rogers said he could look at a lake and pick out the "living rooms," where most of the snapping turtles would dwell and the choice spots where a big dominant Dave Butz–type would have its bailiwick. (Among his observations that science has not tested, Rogers believes that snapping turtles are territorial and observe a pecking order, and also that they may form monogamous pairs.) Within their living rooms, he said, the snapping turtles revealed themselves to anyone who knew their habits: When they lumbered along the bottom, the bubbles from the disturbed air pockets came up in twos, about as far apart as the turtle's front legs. You could calculate how far ahead of the bubbles the turtle would be, and then reach a gaff down even six feet deep and bring it up. If the water was covered with algae, you could look for the discolored trail where the turtle's shell had disturbed the surface; the turtle would probably be right about where the trail ended. Sometimes you could see the lily pads being pulled down, *pffft-pffft-pffft,* or you could study the shoreline for signs of water lapping around rocks. The rocks might be the shells of snapping turtles, which gently rise and fall when they bask near the surface.

For years, a dentist friend had been after him to reveal some of these secrets and finally, Rogers said, he'd acquiesced in a moment of weakness. They went out on a pond in the Rhode Island town where the dentist lived and worked but had never seen any snapping turtles. Almost immediately, Rogers pointed out a half-dozen turtles "shelling" in the distance. " 'You see 'em?' I said. And he said, 'Nope,' so I paddled closer. 'You see 'em?' 'Nope.' I moved in on one quietly and he still didn't see it. 'Well if you put your hand in the water, he could bite it,' I said. I reached down and pulled the turtle out and it was forty-six pounds. He said, 'Oh, my God.' "

(I spoke to the dentist afterward, and he said Rogers took twelve hundred pounds out of the pond that morning, without ever using his call. The dentist said he couldn't help but talk about what he'd seen, and Rogers "almost had a fit afterwards when he thought about what he'd done," exposing himself and his turtles. They are still friends, but Rogers no longer takes anyone turtling.)

Rogers stood up to adjust the stove. He was through the mink and into the muskrat now, and he wanted the temperature right for drying the skins. Nooney had muttered himself to sleep in a corner of the room. Rogers came back, ducking under a chin-high crossbeam where he'd run out of energy for building the cabin wall any higher.

"I know how to call 'em," he said, taking up the skinning knife again. "It was something I heard 'em do one time, and I had a brainstorm." The call, he said, was a device weighing ten or fifteen pounds. "It's nothing like what they hear in nature," he conceded. "But if you were in the woods and you hadn't seen a woman for three weeks, and then you saw a beautiful naked woman run by with blue hair, you'd still chase her, I think." This, at least, was the effect the call had on snapping turtles.

"I had eighteen at one time coming toward me across the flats at Selkirk Shores on Lake Ontario," Rogers said. "Most of the time I get two or three moving. I call them into the shallow water. I just get 'em to reveal themselves and then I go get 'em."

I'd heard that he actually lowered the call into the water, but Rogers decided that he'd said enough. "As long as I know what I know the snapping turtle will always be safe," he said. "If somebody developed a chemical to put in the water that killed all the snapping turtles, he'd live as long as it took for me to get to him and tear out his windpipe."

This affinity with snapping turtles was part of the legend Rogers had acquired—not just that he could call them, but that he could wade in among them unharmed, as if he were one of them, as if the turtles recognized that he had a streak of ferocity in common with them. Once, in Palmyra, Maine, he dumped a twenty-eight-hundred-pound load of snapping turtles into a shallow pond, to clean them up for market. He waded in among them, barefoot and wearing only a bathing suit, and was hauling them out by the tail and even by the head ("Just make sure you hold the mouth shut"), when a group of local boys showed up with the short-lived idea of swimming. "They thought I was catching them that way," Rogers said, and thus are legends born. "People think I'm nuts standing in turtles up to my hips," he noted, but the truth is that snapping turtles tend to be placid in the water.

I asked Rogers if this meant he had never actually been bitten. He put down his knife and started counting marks on his right hand. "One, two, three, four, five, six, seven, eight. Most of the time, though, I get bitten on the sides of the legs. I put fifty or sixty pounds in a bag and I've got to carry 'em out to the truck. They reach their heads out and nip me around the knees. It's worse to get bitten when it's cool because they'll hang on. When it's warm they generally let go." He indicated a fingernail where a snapper once bit to the bone. "Like a knife through butter. It was dark out. I wait till dark so I don't sweat so much when I carry the turtles out. I held my hand up in the moonlight and I could see the blood running down."

Like the turtles themselves, he also had a scar on the back of his neck. "I was riding down along Cayuga Lake with my girl-friend. I have a club cab on my pickup, and I was using old bags for the turtles then. The turtles used to get loose and come through the club cab window. They always went for my window. They'd put a paw on my left shoulder. I got to be nonchalant about it. I'd turn and there'd be a turtle head there, so I'd get a rag and push it back. But my girlfriend saw it and when she screamed I hit the brakes and the turtle got the back of my neck. Another time I hit the brakes and one came tumbling down between my legs, snapping vigorously. It got me on the leg." He shrugged all this off. "I've lost more blood over the years to broken glass than anything."

By now it was past midnight and the work was mostly done. Rogers was planning to head out again around 4:00 A.M. to climb along the Appalachian Trail and do some marten trapping. He didn't need much sleep at this time of year, he said. In another week or two he would head into his own hibernation, at a farm he owned in New York State, where his girlfriend was waiting. He planned to spend the winter working on a novel, about a developer who wanted to turn a wild valley into a suburb; the hero was an outdoorsman who burned down the new houses before anybody could move in. "I hate the suburbs," he said, by way of explanation. "I hate them with a passion."

He would not start turtling again until April, when the snappers came out of their hibernation, and from then until September he'd sleep in the pickup truck, with his feet out the window and the turtles banging around in the back. "When I first started trapping turtles, it was because the man gave me a dollar and that was a day's pay," he said. "It was just an ugly, smelly animal and it was kind of a thrill for a kid. Now, I know there's no one else who likes 'em as much as I do. When I wake up at night, I can tell what time it is by how the turtles are. The hissing and the scratching and the sound of their claws, and the shells rubbing together. That's how close I get to them. I feel I could come back as a snapping turtle and be happy. I'd like to come back as a snapping turtle."

It occurred to me that he would make a good Bosikado.

He grinned: "I'd go and grab suburbanites and drown 'em."

This steadily unfolding sympathy with the snapping turtle had produced certain complications for him. He began to chew around the universal two-o'clock-in-the-morning sensation of being embarked on a deeply flawed but unalterable course through life: "I spend the whole week with turtles. It's just me and them, me and the snappers against the world. And then what do I do? I take 'em into the cities and the suburbs and I sell 'em. I'm selling my friends. Every Thursday night I spend selling, and it's always the same feeling. At the end of the night, the truck's going to be empty and I'll be alone again."

He had one consoling thought in all this. Over the years he had learned to incubate snapper eggs and raise the hatchlings to a good, healthy four pounds. With this stock, he had seeded rivers, refuges, lakes, reflecting pools, water hazards, and ornamental ponds in cities and suburbs and open country all across his range. He was a sort of modern-day John Chapman, the Massachusetts man who spread apple orchards across the Midwest and entered folklore as Johnny Appleseed. But what Rogers was doing had both a suggestion of penance, for his transgressions against the Great Turtle, and also a note of gleeful malice, as if he were subverting the sterile world of the suburbs. "People build ponds," he said. "If it's a good pond, I'll put turtles in it." Three hundred and two of them, in one case, though a half-dozen was more common. Overall, he said, he had stocked a quarter-million snappers, a factor in snapping turtle dispersal that the scientists have doubtless overlooked.

He sat back and showed his big, perfect white teeth. "Under cover of darkness," he said, "you can go anywhere."

Epilogue: Black Hound

The sun was shining over the Atlantic when I arrived at the end of what had been a wet, stormy September day. I was staying the night at a hostel on Crohy Head about five miles from Dunglow, County Donegal. Inside, there was a turf fire and tea, and I had already put in many miles on foot. But the sun drew me out of doors again as soon as I had warmed myself a bit.

The hostel stood on a high spot above pastures running down to the rocks and the water. Clambering over a stone wall, I walked slowly down through the grass. A stream nearby made a sound like schoolchildren murmuring, but the overgrowth hid it completely. Distracted by the water, I didn't notice until the last moment the black dog racing down on me. It ran with a crazy, bucketing speed, and it leaped up at me almost before I had spotted it.

I have always been good with dogs, and I found I was particularly compatible with those in the west of Ireland. It was easy. The dogs there generally look alike and share the same quiet temperament; they are bred to herd sheep. "You seldom get a cross

Sheepdog Leaping

sheepdog," a Mayo farmer had told me, as his dog rested the top of its head against my knee. I hadn't met a cross one yet, and so I didn't panic when the dog came at me.

It twisted in midair, landed at my feet, and was up again immediately, leaping and turning. Like a sheepdog, it didn't growl or bark, but I'd never seen one so wild and frenetic. Most of them I had come across in Ireland were content to sit listening and watching in the corner of a field or to wait patiently in a farmhouse for their meal of potato skins and milk. I saw by the mad, happy look in this dog's eyes that it wanted a tussle. I offered it my arm. Then I wrestled and boxed with it and spun it around, letting it leap for my hand and the sleeve of my jacket. It was skinny and stood only knee-high, so I was able to toss it about without much effort.

The sheepdog started playing other games with me, trying to make me run. It charged me, cutting and bobbing. It dashed down the hill and then came back a bit to tease me on, leaping out of reach each time I drew near. After running tight circles around me, it bolted out of sight, only to appear a few moments later on a knoll two fields away. Its speed and energy startled me. It didn't leap the stone walls with the compact forward motion of a horse. Instead, it took them with an insane springiness that seemed to leave it suspended in the red sunlight high above the wall.

I followed the dog across the fields, playing the game. When I stopped to catch my breath by the ruins of a church, the dog darted in and out of the vacant doors and windows, calling me back to the chase. It drew me on insistently until we got to the bottom of the field, which surmounted a sheer, thirty-foot cliff.

A finger of the Atlantic ran beneath the cliff, and the sea whirled and spat and washed around the rocks below. I paused to take in the scene. The sun, a great red ball, was easing down into the sea. It lit up the red-and-brown cliff face, the rich green grass, the gray stone ruins. Everything in sight was timeless, as if it had not changed in the centuries since the Spanish Armada smashed to pieces on these shores.

But the dog had no interest in ruins. It ran down the irregular rocks to the sea like a house dog going down the back stairs. It was as if the dog could have done it blindfolded. It stopped only to look at me curiously. Twice it came back up onto the grassy headland to urge me on. I didn't move at first. I didn't trust the way the sea rolled into the little inlet and then went sucking back out. Was it a rising or a falling tide? How deep was the water around the rocks?

The dog ran down to the water's edge and continued on, stepping from one rock to another, into the middle of the inlet, where the sea was all around it. It made the trip slowly, looking over its shoulder as if to instruct me. Then once again it returned to me on the headland. There was really nothing to be afraid of, I thought. The weather was fine. There was still plenty of daylight. The dog would be my guide.

I climbed down slowly, unsure of my footing. The dog ran ahead with its characteristic abandon, stopping at intervals for me to catch up. I slipped and fell once, and it came back to inspect me with its gray-blue eyes. It pronounced me well and then stood behind, urging me on.

I looked again at the swirling water and at the clear sky. I thought of Ireland's weather, "changeable as a baby's bottom." Only that afternoon, on the Bloody Foreland, the country's northwestern tip, I had watched this same sea lashing at the rocky coast in a great, white, boiling fury. A blinding rain had whipped in over the low farmland, tearing at the hand-stacked oat ricks. "There's nothing from here till you get to America," one fellow had told me, "so the sea has plenty of room to work up a rage."

From its perch on a rock a few feet away, the dog looked back at me expectantly. The red sun showed it handsomely, all black, with those penetrating eyes. It waited for me. I wondered, only half jokingly, if it was some agent of the sea gods, leading me away from the safety of land. I had been told of demon cats in Irish folklore. Were there demon dogs, too? In the un-Anglicized Irish, my family name (O Conduibh) means "Son of a Black Hound." Was this dog the agent of some personal Irish destiny, some retribution due my ancestors who had lived in the west of Ireland long ago?

I stepped out onto the rock. The dog turned, and with that characteristic springiness, leaped one rock farther out. Then it turned and waited. When I hesitated too long, it came back to draw me on. And so I followed it from rock to rock. It wasn't satisfied until

I had come to the last rock in the little archipelago. We stopped, with the sea rushing around us and cliffs on either side, and at last it sat down like a sheepdog, staring quietly straight ahead.

I followed its gaze, peering anxiously into the west for a squall rolling in from America, but the Atlantic had taken on a glassy evening calm.

The dog was watching the sunset.

I was puzzled and maybe also a little disappointed. Much as I feared it, part of me was actually longing for something other-worldly, even demonic, in the animal world. I had devoted myself to the plain facts of biology and evolution, and yet I still wanted some larger, mythic connection. Maybe it was just a Pleistocene memory, another holdover from our hunter-gatherer history. I wanted to join the dance of fear, worship, curiosity, delight in which the lives of people and animals twine endlessly together. And here this damned dog was handing me—what?—a postcard.

And yet it was a perfect spot. At the mouth of the inlet there was no distraction, nothing to obstruct our view. I settled down and, after a time, accepted that maybe this *was* the dance, or as much of it as I would ever know: a couple of odd creatures side by side on a rock traveling through space and warming them-selves by a distant fire. The sun lit up a great red path straight across the sea. We watched silently as it settled down into the ocean. Then, in the dusk, the dog led me slowly back over the rocks and up onto the headland. Someone was calling when we reached the road. The dog's ears went up, and we parted as abruptly as we had met.

BIBLIOGRAPHY

INTRODUCTION: HEALTHY TERROR

Campbell, J. *The Way of the Animal Powers—Historical Atlas of World Mythology*, Vol 1. New York: Harper & Row, 1983.

Heaney, S. "The Skunk," in *The Norton Anthology of English Literature*, ed. by M. H. Abrams. New York: Norton, 1986.

Huxley, A. *Do What You Will: Twelve Essays*. London: Chatto & Windus, 1949.

Kellert, S. R., and E. O. Wilson, eds. *The Biophilia Hypothesis*. Washington, D.C.: Island Press, 1993.

Pliny the Elder. *Naturalis historia*. With an English translation by H. Rackham. Cambridge, Mass.: Harvard University Press, 1962–1975.

Solinus, G. J. *The Excellent and pleasant worke, Collectanea rerum memorabilium of Gaius Julius Solinus*, translated from the Latin by A. Golding. Delmar, N.Y.: Scholars' Facsimiles & Reprints, 1955.

White, T. H. *The Bestiary: A Book of Beasts*. New York: Putnam, 1960.

Wilson, E. O. *Biophilia*. Cambridge, Mass.: Harvard University Press, 1984.

DAYS OF TORPOR, NIGHTS OF SLOTH

Beebe, W. *Jungle Days*. New York: Putnam, 1925.

———. "The Three-Toed Sloth Bradypus cuculliger cuculliger Wagler." *Zoologica* 7 (1926): 1–67.

Goffart, M. *Function and Form in the Sloth*. New York: Pergamon, 1971.

Montgomery, G. Gene, ed. *The Ecology of Arboreal Folivores*. Washington, D.C.: Smithsonian Institution Press, 1978.

Montgomery, G. Gene, and M. E. Sunquist. "Impact of Sloths on Neotropical Forest Energy Flow and Nutrient Cycling," in *Tropical Ecological Systems—Trends in Terrestrial and Aquatic Research,* ed. by Frank B. Golley. New York: Springer-Verlag, 1975.

Sunquist, M., and G. G. Montgomery. "Activity Patterns and Rates of Movement of Two-Toed and Three-Toed Sloths (*Colopeus hoffmanni and Bradypus infuscatus*)." *Journal of Mammalogy* 54 (1973): 946–954.

Waage, J. K., and G. G. Montgomery. "*Cryptoses choloepi*: A Coprophagous Moth That Lives on a Sloth." *Science* 193 (1976): 157–158.

Wetzel, Ralph M. "Systematics, Distribution, Ecology, and Conservation of South American Edentates," in *Mammalian Biology in South America,* ed. by Michael Mares and H. M. Genoway. Linesville, Pa.: Pymatuning Laboratory of Ecology, University of Pittsburgh, 1982.

THE DEVILBIRD OF NONSUCH*

Beebe, William. *Nonsuch: Land of Water.* New York: Blue Ribbon Books, 1932.

Lefroy, J. H. *Memorials of the Bermudas.* London, 1876. Reprinted London: Eyre & Spottiswoode Ltd., 1932.

Moorehead, A. *Fatal Impact: An Account of the Invasion of the South Pacific, 1767–1840.* New York: Harper & Row, 1966. Reprinted Honolulu: Mutual Pub. Co., 1989.

Verrill, Addison E. *The Bermuda Islands.* New Haven: Yale University Press, 1902.

Wingate, D. "The Restoration of an Island Ecology." *Whole Earth Review Magazine* (Fall 1988): 42–57.

———. "Successful Reintroduction of the Yellow-Crowned Night-Heron as a Nesting Resident on Bermuda." *Colonial Waterbirds* 5 (1982): 104–115.

*David Wingate's work is supported by The Cahow Account, The Bermuda Audubon Society, P.O. Box 1328, Hamilton HMFX, Bermuda.

THE CAVE OF THE BATS

Griffin, D. R. *Echoes of Bats and Men.* New York: Anchor Books, 1959.

———. *Listening in the Dark: The Acoustic Orientation of Bats and Men.* New Haven: Yale University Press, 1958. Reprinted 1986, Ithaca, N.Y.: Cornell University Press.

Hicks, A. "In Search of Wintering Bats." *The Conservationist* (1987): 14–17, 51, 56.

Hitchcock, H. B. "Bat Caves." *Vermont Life* (1962): 26–32.

Kunz, T. H., and E. L. P. Anthony. "Age Estimation and Post-Natal Growth in the Bat *Myotis Lucifugus.*" *Journal of Mammalogy* 63 (1982): 23–32.

Novacek, M. J. "Navigators of the Night." *Natural History* 97 (Oct. 1988): 66–71.

GOOD SCENTS

Brey, Catherine F., and L. F. Reed. *The Complete Bloodhound.* New York: Howell Book House, 1978.

Brisbin, I. L., and S. N. Austad. "Testing the Individual Odour Theory of Canine Olfaction." *Animal Behaviour,* 42 (1991): 63–69.

Leonard, R. M. *The Dog in British Poetry.* London: D. Nutt, 1893.

Rosebury, T. *Life on Man.* New York: Viking, 1969.

Stowe, H. B. *Uncle Tom's Cabin; or Life Among the Lowly.* New York: J. P. Jewett, 1852.

Thurber, J. *Thurber's Dogs: A Collection of the Master's Dogs, Written and Drawn, Real and Imaginary, Living and Long Ago.* New York: Simon & Schuster, 1955.

Tolhurst, W. D., and L. F. Reed. *Manhunters!* Puyallup, Wash.: Hound Dog Press, 1984.

Looking for Mr. Griz*

Abbey, E. *Hayduke Lives!* New York: Little, Brown, 1991.

———. *The Monkey Wrench Gang.* Philadelphia: Lippincott, 1975.

Brown, G. *The Great American Bear Almanac.* New York: Lyons & Burford, 1993.

Craighead, F. C., Jr. *Track of the Grizzly.* San Francisco: Sierra Club Books, 1979.

Gibbs-French, M., and S. French, eds. *The Yellowstone Grizzly Journal.* Evanston, Wyo.: Yellowstone Grizzly Foundation, (periodical) various issues.

McNamee, T. *The Grizzly Bear.* New York: Knopf, 1984.

Peacock, D. *Grizzly Years: In Search of the American Wilderness.* New York: Henry Holt, 1990.

*The Frenches' work is supported by the Yellowstone Grizzly Foundation, 581 Expedition Drive, Evanston, WY 82930.

What's Nice? Mice.

Berry, R. J., ed. *Biology of the House Mouse.* New York: Academic Press, 1981.

Boursot, P., et al. "Origin and Radiation of the House Mouse: Mitochondrial DNA Phylogeny." *J. Evol. Biol.* 9 (1996): 391–415.

Crowcraft, Peter. *Mice All Over.* London: G. T. Foulis, 1966.

Din, W., et al. "Origin and Radiation of the House Mouse: Clues From Nuclear Genes." *J. Evol. Biol.* 9 (1996): 519–539.

Marshall, J. T., Jr. "Systematics of the Genus *Mus*." in *Current Topics in Microbiology and Immunology, Vol. 127.* Heidelberg: Springer-Verlag, 1986.

Perrigo, G. H., and F. S. vom Saal. "Mating-induced Regulation of Infanticide in Male Mice: Fetal Programming of a Unique Stimulus-Response," in *Ethoexperimental Analysis of Behaviour,* ed. by R. J. Blanchard et al. London: NATO Advanced Study Institute Series, 1988.

Sage, R. D. "Wild Mice," in *The Mouse in Biomedical Research,* Vol. 1, ed. by Henry L. Foster et al. New York: Academic, 1981.

CORMORANT HEAVEN

Glahn, J. F., et al. "Food Habits of Double-Crested Cormorants Wintering in the Delta Region of Mississippi." *Colonial Waterbirds* 18 (1995): 158–187.

Grzimek, B. *Grzimek's Animal Life Encyclopedia.* New York: Int'l Specialized Book Service, 1972.

Gudger, E. W. "Fishing with the Cormorant." *American Naturalist* 60 (1926): 5–41.

Lewis, H. F. *The Natural History of the Double-Crested Cormorant* (Phalacrocorax auritus auritus). Ithaca, N.Y. (doctoral thesis), 1929.

Ludwig, J. P. "Decline, Resurgence, and Population Dynamics of Michigan and Great Lakes Double-Crested Cormorants." *Jack Pine Warbler,* Mich. Audubon Soc. 62 (1990): 91–102.

Rogan, W. J., et al. "Congenital Poisoning by Polychlorinated Biphenyls and Their Contaminants in Taiwan." *Science* 241 (July 15, 1988): 334–336.

ACTING LIKE ANIMALS

Edelson, E. *Great Animals of the Movies.* New York: Doubleday, 1990.

Rothel, D. *The Great Show Business Animals.* New York: A. S. Barnes, 1980.

A MOUSE LIKE A SPEAR

Hughes, T. *Collected Animal Poems.* London: Faber and Faber, 1984.

King, C. *The Natural History of Weasels and Stoats.* Ithaca, N.Y.: Cornell University Press, 1989.

———. "Weasel Roulette." *Natural History* (Nov. 1991): 35–41.

Nowak, R. M., and J. L. Paradiso, eds. *Walker's Mammals of the World* (2 vols.). Baltimore: Johns Hopkins University Press, 1983.

Sandell, M. "Stop-and-Go Stoats." *Natural History* (June 1988): 55–65.

Svendsen, G. E. "Weasels," in *Wild Mammals of North America,* ed. by J. A. Chapman and G. A. Feldhamer. Baltimore: Johns Hopkins University Press, 1982.

SHARKS (GREAT WHITE HUNTER AND
HOW SHARKS GOT INTO SUCH DEEP SOUP)*

Benchley, Peter. *Jaws.* Garden City, N.Y.: Doubleday, 1974.

Castro, J. I. *The Sharks of North American Waters.* College Station, Tex.: Texas A&M University Press, 1983.

Ellis, R., and J. E. McCosker. *Great White Shark.* Stanford, Calif.: Stanford University Press, 1991.

Gruber, S. H., ed. *Discovering Sharks.* Highlands, N.J.: American Littoral Society, 1991.

Klimley, A. P., and D. G. Ainley, eds. *Great White Sharks: The Biology of Carcharodon Carcharias.* New York: Academic, 1996.

Manire, C., and S. H. Gruber. "Many Sharks May Be Headed Toward Extinction." *Conservation Biology* (March 1990).

Springer, V. J., and J. P. Gold. *Sharks in Question: The Smithsonian Answer Book*. Washington, D.C.: Smithsonian Institution Press, 1989.

*A leading lobbyist on behalf of sharks is the Center for Marine Conservation, 1725 DeSales St., NW, Washington, DC 20036.

A PORCUPINE WOULD RATHER BE LEFT ALONE

Costello, D. F. *The World of the Porcupine*. Philadelphia: Lippincott, 1966.

Dodge, W. E. "Porcupine," in *Wild Mammals of North America—Biology, Management, Economics,* ed. by J. A. Chapman and G. A. Feldhamer. Baltimore: Johns Hopkins University Press, 1982.

NOTES FROM THE UNDERGROUND

Catania, C. C., and J. H. Kaas. "The Unusual Nose and Brain of the Star-Nosed Mole." *BioScience* 46 (1996): 578–586.

Folitarek, S. "The Distribution and Biology of the Mole (*Talpe europaea brauneri* Satun.) and Mole Catching in the Ukraine." *Bull. Soc. Nat. Moscou Biol.* 41 (1932): 235–302. (In Russian. Translation by J. D. Jackson, 1947, for the Bureau of Animal Population, Oxford University, England. TRANS. 213. F 1712 H.)

Godfrey, G. K., and P. Crowcroft. *The Life of the Mole*. London: Museum Press, 1960.

Gorman, M. L., and R. D. Stone. *The Natural History of Moles*. Ithaca, N.Y.: Cornell University Press, 1990.

Gould, E., W. McShea, and T. Grand. "Function of the Star in the Star-Nosed Mole, *Condylura cristata*." *Journal of Mammology* 74 (1993): 108–116.

Grahame, Kenneth. *The Wind in the Willows*. New York: Scribner's Sons, 1908.

Nowak, R. M., and J. L. Paradiso, eds. *Walker's Mammals of the World* (2 vols.). Baltimore: Johns Hopkins University Press, 1983.

Yates, T. L. "The Mole That Keeps Its Nose Clean." *Natural History* 92 (Nov. 1983): 55–61.

Yates, T. L., and R. J. Pedersen. "Moles," in *Wild Mammals of North America,* ed. by J. A. Chapman and G. A. Feldhamer. Baltimore: Johns Hopkins University Press, 1982.

JUNGLE DAYS*

Emmons, L. H. *Neotropical Rainforest Mammals*. Chicago: University of Chicago Press, 1990.

Gentry, A. H. *Four Neotropical Rainforests*. New Haven: Yale University Press, 1990.

————. *Woody Plants of Northwest South America (Colombia, Ecuador, Peru), with Supplementary Notes on Herbacious Taxa*. Chicago: University of Chicago Press, 1993.

Stevens, W. K. "Biologists' Deaths Set Back Plans to Assess Tropical Forests." *New York Times* (Aug. 17, 1993): C1, C9.

Stotz, D. F., et al. Neotropical Birds: Ecology and Conservation. Chicago: University of Chicago Press, 1996.

*The RAP Team's work is supported in part by the Parker-Gentry Fund, Conservation International, 1015 18th St., NW, Suite 1000, Washington, DC 20036.

SLEEPING WITH SNAPPING TURTLES

Congdon, J. D., et al. "Ontogenetic Changes in Habitat Use by Juvenile Turtles, *Chelydra serpentina* and *Chrysemys picta.*" *Canadian Field-Naturalist* 106 (1992): 241–248.

Coulter, M. W. "Predation by Snapping Turtles upon Aquatic

Birds in Maine Marshes." *Journal of Wildlife Management* 21 (1957): 17–21.

Ernst, C., R. W. Barbour, and J. E. Lovich. *Turtles of the United States and Canada.* Washington, D.C.: Smithsonian Institution Press, 1994.

Kiviat, E. "A Hudson River Tidemarsh Snapping Turtle Population." *Trans. Northeast Section, Wildlife Soc.* 37 (1980): 158–168.

Seton, E. T. *Rolf in the Woods.* New York: Grosset & Dunlap, 1911.

INDEX

Page numbers in *italics* refer to illustrations.

Index